Dynamics of Offshore Structures

Dynamics of Offshore Structures

Editor

Rajeev Dubey

Dynamics of Offshore Structures
Edited by **Rajeev Dubey**

Printed in 2017

ISBN: 978-1-68117-358-0

Library of Congress Control Number: 2015941550

© 2016 by
SCITUS Academics LLC,
616, Corporate Way, Suite 2, 4766,
Valley Cottage, NY 10989

www.scitusacademics.com

This book contains information obtained from highly regarded resources. Copyright for individual articles remains with the authors as indicated. All chapters are distributed under the terms of the Creative Commons Attribution License, which permits unrestricted use, distribution, and reproduction in any medium, provided the original author and source are credited.

Notice

Reasonable efforts have been made to publish reliable data and views articulated in the chapters are those of the individual contributors, and not necessarily those of the editors or publishers. Editors or publishers are not responsible for the accuracy of the information in the published chapters or consequences of their use. The publisher believes no responsibility for any damage or grievance to the persons or property arising out of the use of any materials, instructions, methods or thoughts in the book. The editors and the publisher have attempted to trace the copyright holders of all material reproduced in this publication and apologize to copyright holders if permission has not been obtained. If any copyright holder has not been acknowledged, please write to us so we may rectify.

Contents

Preface vii

Chapter 1   Plastic Optical Fibre Sensors for Structural Health
            Monitoring: A Review of Recent Progress 1

            K. S. C. Kuang, S. T. Quek, C. G. Koh, W. J. Cantwell,
            and P. J. Scully

Chapter 2   Investigation of Wave-Structure Interaction Using
            State of the Art CFD Techniques 39

            Jan Westphalen, Deborah M. Greaves, Alison Raby,
            Zheng Zheng Hu, Derek M. Causon, Clive G. Mingham,
            Pourya Omidvar, Peter K. Stansby, and Benedict D. Rogers

Chapter 3   Advantages of the Green Solid State FSW over
            the Conventional GMAW Process 93

            Hasan I. Dawood, Kahtan S. Mohammed, and Mumtaz Y. Rajab

Chapter 4   Corrosion Behavior of Carbon Steel in
            Synthetically Produced Oil Field Seawater 123

            Subir Paul, Anjan Pattanayak, and Sujit K. Guchhait

Chapter 5   Hydrodynamics of an Inclined Gas–Liquid
            Cocurrent Up flow Packed Bed 149

            Hana Bouteldja, Mohsen Hamidipour, and
            Faïçal Larachi

Chapter 6   Interaction between Aqueous Solutions of
            Hydrophobically Associating Polyacrylamide and
            Dodecyl Dimethyl Betaine 171

            Zhongbin Ye, Guangfan Guo, Hong Chen, and Zheng Shu

Chapter 7   Opportunities and Challenges pf Robotics and
            Automation on Offshore Oil & Gas Industry 197

            Heping Chen, Samuel Stavinoha, Michael
            Walker, BiaoZhang, and Thomas Fuhlbrigge

Chapter 8	**Vibration of Slender Structures Subjected to Axial Flow or Axially Towed in Quiescent Fluid**..............................219	
	L. Wang and Q. Ni	
Chapter 9	**Water-Air CO_2 Fluxes in the Tagus Estuary Plume (Portugal) during Two Distinct Winter Episodes** 281	
	Ana P Oliveira, Marcos D Mateus, Graça Cabeçadas, and Ramiro Neves	

Citations..315

Index..319

Preface

Dynamics of Offshore Structures provides an integrated treatment of the main subject areas that contribute to the design, construction, installation, and operation of fixed and floating offshore structures. The book begins with an overview of offshore oil and gas development and offshore structures. Separate chapters follow on the ocean environment; basic fluid mechanics; gravity wave theories; fluid loading on offshore structures; hydrostatics and dynamic response of floating bodies; and model testing of offshore structures. Basic fluid mechanics, wave theory, hydrodynamics, naval architecture, and structural analysis to meet the needs of students reading ocean engineering or naval architecture, at both undergraduate and postgraduate levels. Basic equations and theoretical results are derived in a rigorous manner but sections on model testing, full-scale measurements, design, and certification are also induced to ensure that the book is of value to professional engineers seeking a balanced treatment of fundamental and practical issues.

Editor

Chapter 1

Plastic Optical Fibre Sensors for Structural Health Monitoring: A Review of Recent Progress

K. S. C. Kuang[1], S. T. Quek[1], C. G. Koh[1], W. J. Cantwell[2], and P. J. Scully[3]

[1]Department of Civil Engineering, National University of Singapore, Block E1A, #07-03, 1 Engineering Drive 2, Singapore 177576

[2]Department of Engineering, University of Liverpool, Brownlow Hill, L69 3GH Liverpool, UK

[3]The Photon Science Institute, University of Manchester, Oxford Road, M13 9PL Manchester, UK

ABSTRACT

While a number of literature reviews have been published in recent times on the applications of optical fibre sensors in smart structures research, these have mainly focused on the use of conventional glass-based fibres. The availability of inexpensive, rugged, and large-core plastic-based optical fibres has resulted in growing interest amongst researchers in their use as low-cost sensors in a variety of areas including chemical sensing, biomedicine, and the measurement of a range of physical parameters. The sensing principles used in plastic optical fibres are often similar to those developed in glass-based fibres, but the advantages associated with plastic fibres render them attractive as an alternative to conventional glass fibres, and their ability to detect and measure physical parameters such as strain, stress, load, temperature, displacement, and pressure makes them suitable for structural health monitoring (SHM) applications. Increasingly their applications as sensors in the field of structural engineering are being studied and reported in literature. This article will provide a concise review of the applications of plastic optical fibre sensors for monitoring the integrity of engineering structures in the context of SHM.

INTRODUCTION

In recent years, structural health monitoring has attracted significant interest from academia, government agencies, and industries involved in a diverse field of disciplines including civil, marine, mechanical, military, aerospace, power generation, offshore and oil and gas. The aim of SHM is to detect damage initiation and subsequently monitor the development of this damage using structurally-integrated sensors in order to provide early warning and other useful information for successful intervention to preserve the structural integrity of the host. A number of commonly monitored parameters used for SHM applications include the detection or measurement of strain, load, displacement, impact, pH-level, moisture, crack width, vibration signatures, and presence of cracks.

Over the last two decades, optical fibre sensors have attracted substantial attention and shown to be capable of monitoring a wide range of physical measurands for SHM applications. The advantages of optical fibre sensing in engineering structures are well known and these include their insensitivity to electromagnetic radiation (especially in the vicinity of power generators in construction sites), being spark-free, intrinsically safe, non-conductive and lightweight, and also their suitability for embedding into structures. To date, a number of key optical fibre sensors have been reported and their applications for damage detection in composite structures are given in review articles elsewhere [1, 2]. Optical fibre-based sensors such as fibre Bragg gratings (FBG), intensiometric and polarimetric-type sensors and those based on interferometric principles (e.g., Fabry-Perot) have been shown to offer specific advantages in their niche area of applications.

Of the various types of optical fibre sensors, intensiometric sensors represents one of the earliest and perhaps the most direct and basic type of optical fibre sensor used for SHM purposes [1]. Here, the sensing principle is straightforward and relies on monitoring the intensity level of the optical signal as it modulates in response to the measured quantity. Although monitoring of the intensity level of optical signal has often been cited to be a drawback as a result of possible power fluctuation in the signal level and influence of external environment unrelated to the measured parameter (e.g., micro and macro bending along the fibre length), standard referencing techniques may be used to counter this problem. With the availability of stable and inexpensive light sources and low bend-sensitivity fibres, the intensity-based approach offers excellent commercial prospect for large-scale applications from a cost-effectiveness point of view. In addition, the intensity-based technique is also suitable for frequency analysis in vibration measurements since precise and absolute measurement of the structural strain or displacement values are not required—given that the sensor has sufficient sensitivity to detect the oscillatory nature of the vibration signal. Plastic optical fibres (POFs) with their large core sizes (diameters ranging typically from 0.25 mm to 1 mm are readily available) and high numerical apertures (0.47) lend themselves well to be used as intensity-based optical fibres sensors. Indeed, many of the POF sensors developed and demonstrated for a variety of SHM applications in recent times are based on intensity modulation using these multi-mode fibres [3–16]. The core of the fibre could be made from polymethymethacrylate (PMMA), polycarbonate (PC), polystyrene (PS) and more recently cyclic transparent optical polymer (CYTOP), which offers the lowest attenuation of 50 dB/km at 650 nm compared to 160 dB/km for PMMA-based POF. The cladding layer of the fibre is generally made of fluorinated polymers. At present, most POF sensors are step-index PMMA-based due to their wider availability and lower cost. Other variants of POFs including multicore fibres, double-step-index fibres, multi-step-index fibres and graded-index fibres have also been introduced to improve the bandwidth and to lower the bending sensitivities (by means of multiple smaller cores and optimising on the refractive index profile). Single-mode POFs are presently obtainable commercially (e.g., Paradigm Optics Inc.) although their availability is still limited worldwide. Bragg gratings, which are commonly applied to single-mode silica fibres using ultra violet laser light to create the interference pattern to induce periodic changes in refractive index of the core, have also been demonstrated on doped plastic optical fibres and undoped bulk PMMA in recent years [17–19]. More recently, micro-structured POF have been introduced and these have received significant attention as a promising class of fibre for new sensor applications [20, 21], achieving unique optical properties via a pattern of holes down the full length of the fibre. Optical properties include enabling single-mode fibre to be made from a single matrix material with characteristics controlled by photonic bandgap effects. Unlike single mode POF, single-mode microstructured POF has a visible loss of around 1 dB/m and are single moded for wide (theoretically endless) range of wavelengths. Bragg gratings [22] and long period gratings [23] have been created within mPOF. Advantages include the possibility of optimising the sensitivities of the different loss features to a range of measurands by adjusting the hole geometry, and using asymmetric microstructures for directional bend sensitivity [24].

In addition to being cheaper than their glass-counterpart, plastic fibres offer better fracture resistance and flexibility compared to bare glass fibres. They also offer ease of termination, safe disposability and ease of handling. It has been reported that plastic optical fibre has an elastic limit of 10% compared to 1% in silica and can withstand strains more than 30% without breakage [25]—this could be a significant benefit for structural health monitoring applications involving large strains greater than that measurable by glass-based fibre sensors. For monitoring internal parameters of a structure, for example, when it becomes desirable to embed sensors within concrete structures, POF sensors offer a possible solution since the extremely alkaline (pH 12) environment of the concrete mixture is known to be corrosive to standard glass fibres [26]. Also, the presence of moisture can weaken the glass core and accelerates crack growth in the fibre. For glass-based sensor, although a polymer coating may be applied in order to protect the glass fibre from the corrosive environment, this will incur additional cost. Finally, glass-based optical fibre sensors are fragile and in general not amenable to rough handling and are highly susceptible to fracture in harsh engineering environment. In view of the advantages associated with plastic optical fibres, intensive research is underway to assess their potential for smart structure and structural health monitoring applications.

RECENT DEVELOPMENT IN POF SENSORS FOR SHM

Intensity-Based POF Sensors for SHM

The ease of monitoring the light intensity level in these large core fibres (typically 1 mm step-index multi-mode type) naturally leads to their development as intensity-based sensors. The availability of in-expensive solid-state light emitters and detectors allows the POF sensors to be conveniently integrated to external set-ups such as control and data acquisition systems. Indeed, the simplicity in design associated with intensiometric measurements has resulted in the various applications of POF sensors not only for SHM but for a variety of other sensing applications [27–29]. POF sensors were demonstrated to have the capability to measure parameters such as strain, curvature, bending displacement as well as for detecting cracks within the structure subjected to either quasistatic or dynamic loading [3–16]. In general two classes of intensity-based sensing have been reported and they are grouped based on whether the optical fibre is an intrinsic or extrinsic sensor. In an intrinsic sensor, modulation of the optical signal is a direct result of the physical change in the optical fibre in response to some measurands (e.g., signal change due to the micro- or macro-bending of the fibre). On the other hand, in an extrinsic optical fibre, the signal modulation takes place outside the optical fibre (e.g., signal change due to the changing gap distance between two cleaved fibre surfaces).

Kuang et al. [3] investigated the use of a low-cost, intensity-based intrinsic POF sensor for monitoring the mechanical response of a number of plastic specimens. In their study, the plastic fibre used (ESKA CK40) was a 1 mm diameter multimode step-index type supplied by Mitsubishi Rayon Co. Ltd. By removing a segment of the POF's core and cladding layer over a pre-determined length, the aim was to promote light loss in this region due to reduction in the number of modes undergoing total internal reflection when the fibre was bent. The sensitised region (ranging from 70 mm for smaller specimens to 300 mm for larger ones) was noted to possess directional sensitivity and hence important to ensure the relative planar orientation of the segment of the POF sensor and the direction of loading. The study demonstrated that the POF sensor used exhibited high responsiveness to bending (strain-normalised optical loss coefficient of approximately $1.8 \times 10^{-5}/\mu\varepsilon$) and could be configured to render it sensitive to in-plane axial loads by simply curving the sensing region of the POF in the appropriate orientation with respect to the direction in which the strain is to be measured. Figure 1 (a) shows the signal of POF sensor under a cyclical flexural load while Figure 1(b) illustrates the repeatability of the sensor configured for tensile strain measurement. Although the tests stopped at bending and tension strains of 0.7% and 1.2% respectively, the ability of the POF to measure higher strain values was expected to be achievable. Strain values up to 15.8% have been reported in a single-mode POF by Kiesel et al. [30, 31] while other studies using standard POF have measured strains up to 45%, although it was noted that depending on strain rate and temperature, the fibre could endure more than 80% strain [32, 33].

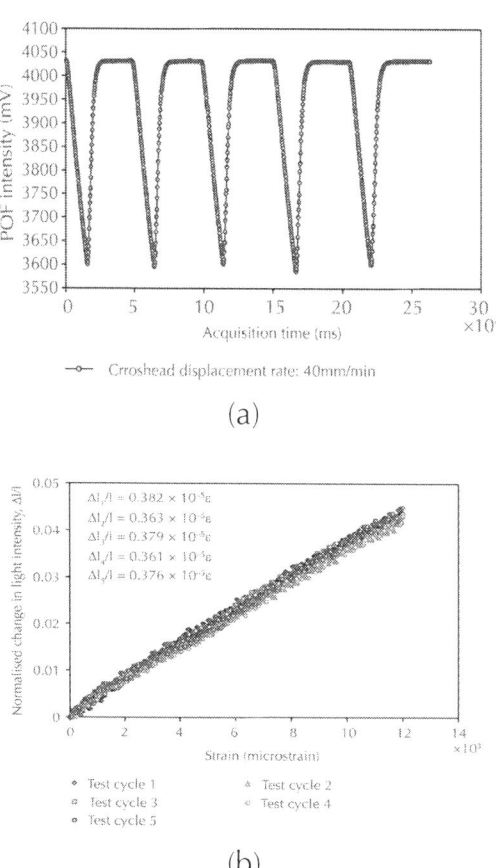

Figure 1: (a) Plot of the POF sensor response under cyclic flexural loading highlighting the stability and responsiveness of the sensor (after [3]). (b) Plot showing the POF sensor signal during a series of tensile tests (after [3]).

The potential use of POF for SHM purposes was also investigated for the dynamic monitoring ability in fibre composites [9, 10]. The upper limit of the frequency tested was 30 Hz (limited by the motor used). Here a POF sensor, identical in terms of theoretical background and operating principle to Kuang et al. [3], was attached to a cantilever-type composite beam to monitor the free vibration of undamaged and damaged specimens following low-velocity impacts. The sensitised POF sensors used was sufficiently sensitive to monitor the change in the damping ratio to characterize the reduction in postimpact flexural modulus and residual strengths of a composite beam with increasing level of impact damage. In the experiment, the POF sensor was able to detect a change in the damping ratio as small as 2.5%. In a later study [34], the POF sensor was applied to a nickel-titanium fibre metal laminate to monitor the morphing response of the hybrid laminate and the POF signal was found to agree well with collocated electrical strain gauges. Following activation of the smart fibre metal laminate (FML) by air through a heat gun, the shape memory alloy (SMA) layers deformed according to the shape it was trained (i.e., curved) and the POF sensor was found to faithfully monitor the flexural response of the smart composite. Figure 2 shows the response of the POF compared to the strain gauge reading.

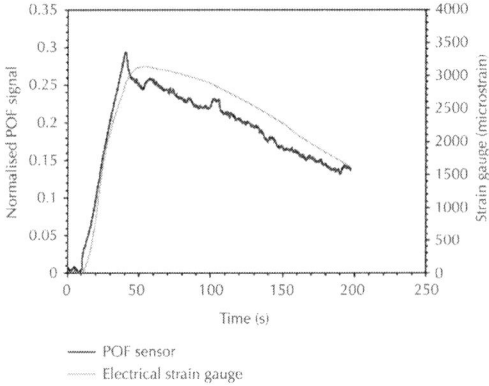

Figure 2: Plot showing the response of the POF sensor following activation of the Ni-Ti sheet (after [34]).

Further work on the use of an extrinsic POF sensor to monitor the deflection of a smart composite was done by Kuang et al. [35]. The operating principle was straightforward—the sensor relied on the monitoring of the optical power transmitted through an air-gap between two cleaved optical fibre surfaces. The two fibres were aligned within a housing in which the fibres could slide smoothly. The gap between the cleaved surfaces changes in proportion to the applied strain resulting in the increase or decrease of the transmitted power. A standard red LED (650 nm) and photodiode were used to illuminate and monitor the optical power during the test. Here, thin-film heating technology was introduced to assess their potential for their integration into the smart FML for SMA activation. To provide the sensor feedback signal, the POF sensor was attached to the smart composite allowing the deflection or morphing response of the specimen to be accurately controlled. The POF sensor used here was modified based on a design described in an earlier work [15]. The desired amount of deflection of the beam was pre-set using a controller and the POF sensor reading was used as a feedback signal to achieve the desired deflection. The deflection of the beam specimen was monitored continuously via a data acquisition set-up which logged the POF sensor output simultaneously. Figure 3 shows the data of the POF sensor readings at three different FML beam deflections (A) to (C). The result shows that the FML beam could be controlled accurately to within 3% of the desired deflection using the POF feedback data with very little overshoot.

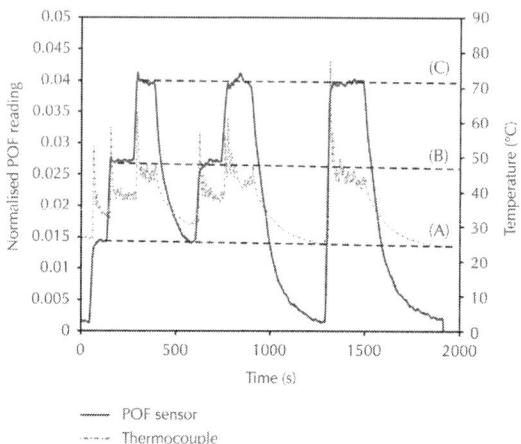

Figure 3: Plot showing the response of the POF sensor highlighting the successful use of the POF data in the control loop to achieve the amount of deflection desired (after [35]).

Another variation of the intensity-based POF sensor for monitoring structural displacement was proposed by Babchenko et al. [4] based on the bending of a multi-looped POF that has sensitised multistructural imperfections on the outer side of its core. The structural imperfections were created on the outer side of the fibre's core by abrading the fibre surface. The imperfections were created in the form of small scratches perpendicular to the curve plane similar in concept to earlier studies [3, 7, and 10]. The increased loss of light at the sensitised region due microbending was then related to the amount of displacement. The study used a simple mechanical set-up where the fibre sensor was located between a top and bottom plate connected to a micrometer allowing the amount of bending of the fibre to be controlled. The authors argued that by adding more loops to the fibres and additional imperfections on the apex of the curve section of the POF, an inexpensive POF sensor could be created to monitor a variety of physical measurands, including strain, stress, vibration and pressure although the authors have not conducted any specific studies to show the actual performance of the proposed sensor to monitor the various loading conditions listed.

POF sensors have been applied to concrete structures in the field of civil engineering in view of their ruggedness and ease of handling compared to glass-based fibre sensors. Kuang et al. [11], conducted a series of flexural tests on scaled-specimens where POF sensors were attached to the bottom surface of the beam and showed that the sensors used were of sufficient sensitivity to detect the presence of hair-line cracks as illustrated in Figure 4. A crack width of approximately 0.04 mm was successfully detected using the POF sensor. In order to improve the sensitivity of the POF to beam deflection and crack initiation, a segment of the POF cross sectional profile was removed over a predetermined length (7 cm for a series of scale-model specimens and 30 cm for full-scale specimens) by abrading the surface of the POF using a razor blade. In principle, the sensitisation process increases the loss in mode propagation when the fibre is bent. Exposing the fibre core by removing the cladding layer to create an evanescent field sensor is a well-known technique commonly exploited for sensing purposes. Following tests on the scaled-specimen, the authors demonstrated the use of POF sensors to monitor the response of three-meter-long concrete beams subjected to a quasistatic lateral load in a three-point bend set-up. In their study, multimode step-index plastic optical fibres were successfully applied to detect initial cracks in the beam and subsequently to monitor post-crack vertical deflection and finally to detect failure cracks in concrete beam. Figure 5 shows the plot summarising the results of the tests carried out for crack detection in concrete specimens.

Figure 4: Photomicrographs showing the intersection of the crack in the crack specimens with the POF sensor (after [11]).

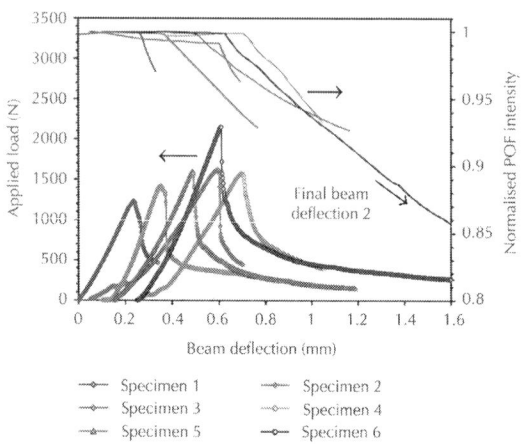

Figure 5: Summary of results for three-point bend tests on concrete specimens (beam deflections for specimens 2–6 have been offset for clarity of presentation) (after [11]).

In another study related to structural health monitoring of concrete beams, a liquid-filled extrinsic POF sensor design was employed to monitor the central deflection of a concrete specimen in a three-point bend configuration [14]. Here, four extrinsic POF sensors were used—each with a different liquid opacity injected into the housing cavity shown schematically in Figure 6(a). The principle of operation is the same as that described earlier where the transmitted optical power across a gap between two cleaved POF surfaces was monitored and related to the applied load or strain. Instead of an air-gap, the addition of an opaque liquid medium in the cavity of the housing increases the strain sensitivity of the POF sensor (up to approximately 25 times) as shown in Figure 6(b). Following the initiation of crack at the bottom surface of the concrete beam, the electrical strain gauge failed instantly while the POF was able to continue monitoring the response of the beam under the transverse load highlighting the advantage of the POF sensors over electrical strain gauges in this particular application. As shown in Figures 7(a) and 7(b), the collocated electrical strain gauge was damaged at the first crack of the beam and was rendered useless limiting its usefulness for structural health monitoring purposes where surface cracks are frequently encountered. The POF sensors, however, did not appear to be significantly affected by the crack and were able to continue monitoring the loading process even after severe crack damage (crack width of approximately 2 mm and a corresponding POF strain of 4.7%) has taken place in there steel reinforced beam specimen. Since the POF sensor was attached to the beam at the two points (which define its gauge length), the propagation of crack across the sensor has insignificant detrimental effect on its measurement capability.

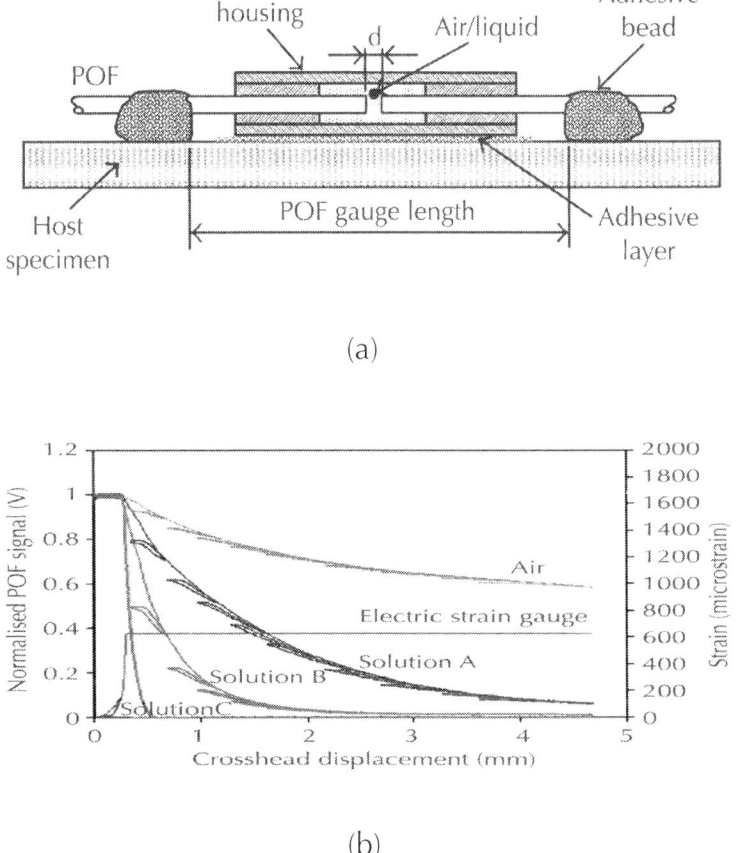

(a)

(b)

Figure 6: (a) Schematic of extrinsic POF sensor. (b) Plot showing the different POF sensor sensitivities corresponding to the different opacity of liquid used and the response of the electrical strain gauge attached to the bottom side of a concrete beam during a quasistatic cyclic test (after [14]).

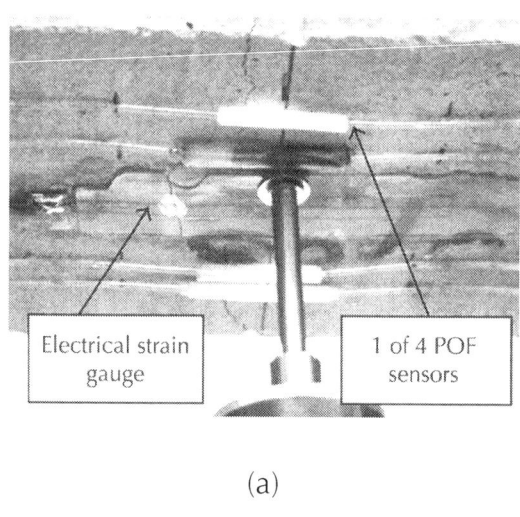

(a)

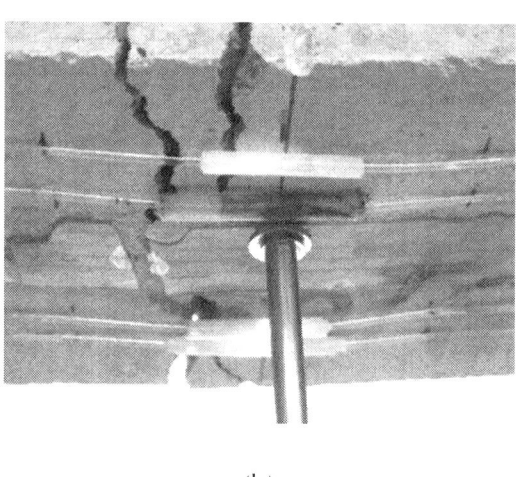

(b)

Figure 7: (a) Photograph showing the crack line across the electrical strain gauge at approximately 0.3 mm beam central deflection. (b) Photograph showing the widening of the crack after several loading cycles—the damage of the electrical strain gauge is evident while the POF sensors continued to monitor the loading process (after [14]).

In view of the potential of the POF sensor for vibration detection, it has also been used in another study on system identification for SHM in a composite beam using genetic algorithm as the parametric search method. An analysis using fast-Fourier transform of the acquired POF vibration signal compared well with a collocated piezofilm sensor and the result showed that the sensor was capable of detecting the shift in the various modal frequencies associated with different system characteristics or damage level [12–14]. The highest frequency detectable in the study was in excess of 1 kHz [12] highlighting the potential of the system for vibration-based structural health monitoring.

In addition to concrete structures, POF sensors have been applied to monitor large strains (defined as greater than 10%) developed in geotextile materials. Kuang et al. [36] reported that the intensity-based sensor used in their work could be customised to monitor strains as high as 40% or more. Based on a previous design [15], the cavity of the housing for this large strain sensor has been readapted and is shown schematically in Figure 8. The basic operating principle of the POF sensor used here relies on measuring the displacement of two cleaved fibre surfaces housed within the tube. Since the two ends of the POF were free to move under an applied axial load, the sensor strain measuring capability was not limited to the yield strain or elastic limit of the POF itself. The authors reported that the signal output of the sensor was directly related the separation of the two end faces and the sensor was initially calibrated with a linear variable displacement transducer before being attached to the geotextile host for strain measurement. POFs are available with protective polyethylene jackets and have a good resistance to damage under marine conditions and therefore are suitable candidates for a marine environment. Being inexpensive to produce and interrogate, it was proposed that the POF sensors were more cost-effective than other optical-based sensors such as fibre Bragg grating (FBG) sensors—for comparison, a POF sensor cost less than US$1 while an acrylate-recoated FBG sensor would cost typically US$100. More significantly, an FBG interrogator could range from US$15,000 to US$40,000 while in contrast; it is possible to fabricate an intensity-based system for under US$200 using off-the-shelf parts highlighting the economic attractiveness of the proposed system.

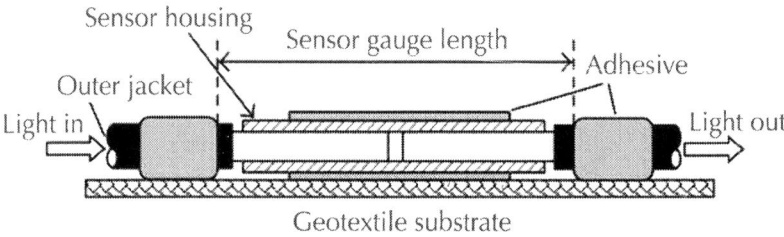

Figure 8: Schematic of POF sensor and location of adhesives (after [36]).

POF has also been embedded in composite materials to monitor damage development. Takeda et al. [5, 6] utilised small diameter multimode POFs (250 μm) for detection of matrix crack in advanced fibre composites. The POFs were embedded in both unidirectional and cross-ply composite lay-up. The sensors were subjected to the same curing cycle as the composites and the authors successfully demonstrated the possibility of embedding POFs into the fibre composites. This approach relies on changes in the optical power-strain relationship to infer damage. It was reported that when the unidirectional specimen was loaded in the axial direction, the optical power decreases linearly with strain while no damage was observed. In contrast, as a result of the transverse crack in the cross-ply laminate specimen, a non-linear optical response was noted following the initiation of cracks in the specimen. Takeda et al. contended that this observation supported their predicted response and hypothesised that the non-linearity observed in the optical response was an indication of local deformation of POF resulting from the damage in the host and hence demonstrated the viability of the technique for detection of transverse damage in composite materials.

Wong et al. [28] embedded a chemically-tapered POF in a carbon fibre composite to examine the potential for strain measurement.

In the study, a POF with taper length of approximately 10 mm was used. Tapering a fibre forms a more sensitive evanescent field sensor because, as the fibre diameter decreases further with strain, the evanescent field penetrates further into the cladding rendering it sensitive to the applied strain. A series of tensile tests was conducted on a composite test beam with an embedded tapered POF sensor. The POF was embedded at the mid-plane of a four-ply woven carbon-fibre epoxy prepreg (Stesapreg EP121-C15-53). The composite lay-up was inserted in a picture frame mould and was processed at 75°C under 3 bars for 8 hours to allow curing of the epoxy matrix. The test specimen was subsequently examined and the POF sensor was found to be capable of withstanding the processing conditions. The authors showed that the embedded POF was capable of monitoring up to 1.4% strain with good repeatability.

Embedding of POF sensors in composite materials requires careful selection of the type of POF material since the processing temperatures vary according to the material system of the composites used. Although certain classes of POF such as polycarbonate and CYTOP are able to sustain higher processing temperatures compared to PMMA POFs without suffering significant optical and structural degradation, the range of composites which could be embedded with POF sensors is clearly limited to the maximum operating temperatures of the chosen POF. Based on the literature reviewed, it is reasonable to conclude that intensity-based POF sensors offer a simple, in-expensive yet effective approach in monitoring specific aspect in structural health monitoring, in particular where high-resolution and precise measurements are not key requirements in the application. The concern over the fluctuation of optical intensity due to macrobending along the fibre or other perturbation not related to the measured quantity could be overcome with careful placement of the fibre (if possible) and referencing techniques and hence would not pose a significant barrier to its deployment. It has also been shown in the above review that, the apparent simplicity in sensor design, signal interrogation and acquisition, intensity-based monitoring approach using standard LEDs, photodetectors and low-cost data acquisition units allows users to implement a working system readily without the need for expensive equipment. The ease of handling large core intensity-based POF sensors outside the laboratory environment adds to its attractiveness for health monitoring of engineering structures. The review also provided examples of the use of intensity-based POF sensors for measurement of strain values less than a few hundred microstrains to that exceeding 40%. The sensors were also demonstrated as surface bonded or embedded in structures. These reports highlight the versatility of the intensity-based POF sensors for measurement commonly performed in engineering applications and would indeed be the preferred technique in certain cases of SHM applications following careful consideration of its limitations and benefits.

OTDR-Based POF Sensors for SHM

Distributed strain sensing using a single fibre enables monitoring over a long section of structure and is highly desirable in structural health monitoring. Optical time-domain reflectometry (OTDR) is a well-known technique in telecommunication for fault analysis and has recently been applied to multi-mode standard POF for strain sensing applications. Despite large fibre core size and hence significant modal dispersion, some success was demonstrated for SHM applications. OTDR sensing exploits the monitoring of the backscatter light in an optical fibre following the launched of a short optical pulse at one end of the fibre. The backscatter signal is recorded as a function of time and then converted to distance measurement. Perturbations, such as strain or defects along the length of the fibre will result in either a peak reflection or loss in the backscatter signal at the location of the perturbation.

Husdi et al. and Nakamura et al. [37, 38] tested two types of PMMA-based multi-mode step-index fibres Eska Premier GH-4001-P and Super Eska SK-40 for the effect of mechanical deformation and temperature on the transmission and reflection properties of these fibres. A commercial photon counting OTDR system (with a position scale resolution of 10 mm) for measuring the very weak backscattering light was used to interrogate the POF sensor. The study showed some interesting effects on the OTDR signal including the response due to (i) a flaw along a fibre (along a 300 m long fibre), (ii) a small bend (bending diameters ranging from 8 mm to 52 mm of a 300 m long fibre), (iii) transverse clamping (clamped over a 50 mm section at a distance 30 m from the input end), (iv) torsional strain (a section of 10 mm was twisted at a distance of 40 m from the input end along a 200 m long cable), (v) axial strain (over a section of 10 cm) and (vi) temperature (a 10 m section at a 300 m long cable was immersed in water of various temperatures). Based on the results, the authors suggested the possibility of discriminating the different types of external perturbation by the specific change in the shapes of the backscattered traces and further detailed investigation will be necessary to correlate the various POF responses to the applied perturbations. The authors argued that if low-cost instrumentation could be developed, the POF-OTDR system will be a competitive candidate for short range distributed SHM.

Fukumoto et al. [39] extended the work on POF-OTDR by conducting a feasibility test of the distributed strain sensor for detecting deformation in wooden structures. In their work, the authors were able to detect the direction and magnitude of deformations at four corners of a rectangular wooden frame. The "memory effect" of the POF was also studied and it was reported that when strain is applied to the POF cable, it could be memorised through the plastic deformation of the core material of the POF, and could be read out using the OTDR even after the strain was removed. The spatial resolution of their set-up was reported to be 5 m for the conventional step-index PMMA POF used (Eska Premier GH-400l-P).

The application of POF using OTDR technology has also been reported elsewhere by Krebber and workers [32, 33, and 40]. In their studies, standard SI PMMA POF was integrated into geotextile materials and was shown to be capable of measuring up to 45% strain. Here, it was observed that the level of the backscattered light increases in a non-linear manner with strain up to 16% at locations where strain was applied to the POF. The results concur with that reported by Husdi et al. [37, 38]. Due to the high loss experienced in standard SI PMMA POF (150 dB/km), perfluorinated graded-index (GI) POFs (loss of 30 dB/km) were also studied for distributed strain sensing. It was observed that the length of fibre monitored extended from 100 m for standard PMMA POF to 500 m in these GI-POFs due to the lower modal dispersion in these graded-index fibres [32, 41]. For the GI-POFs studied, it was highlighted that the GI-POF tested exhibit a rather nonlinear backscatter increase up to about 3% strain, above which no further backscatter increase was observed.

In addition to strain monitoring in geotextile materials, the POF was also used for detection of crack opening in masonry structures up to 20 mm in steps of 2 mm [32]. Two displacement transducers were used as reference sensors which monitored the width of the crack opening continuously. The results showed that the OTDR backscatter signal increased in respond to widening crack width (up to 25 mm) highlighting the feasibility of using POF OTDR sensor to detect cracks in masonry ad concrete structures. The POF sensor was integrated into a geotextile and then surfaced attached to a concrete beam specimen with a small pre-crack. The backscatter signal for two textile specimens was found to increase in a reasonably linear manner (approximately 0.05 dB/mm crack width) although the authors highlighted that further tests will need to be conducted to obtain reproducibility in the results.

Although the technique above is excellent for monitoring long, large sections of structures/materials, it is primarily suited for quasistatic measurements since data acquisition and processing time of a few seconds to a few minutes are required particularly if high resolution measurement of extended POF length is important to the user. The use of OTDR technique for measurement of strain would only be meaningful if the strain levels are of the order of 1% and above and a gauge length of tens of centimetre to 1 m. This would be suitable for very large structures with large strains such as the deployment of the POF-OTDR sensor in geotextiles materials as outlined in the review above but may encounter problems if applied to structures with smaller dimension. In addition, the POF

OTDR equipment may be prohibitively costly in most situations and hence limited to special niche areas in which their distributed strain monitoring capability is exploited and where their initial investment could be justified.

Interferometric-based POF Sensors

Recent progress in the fabrication of single-mode POFs has made possible the use of these fibres for large-strain applications based on interferometric sensing techniques. Here, the principle of operation involves the monitoring of the phase-shift of the propagated light in the test fibre under an applied strain relative to an unstrained reference fibre. The phase-shift is monitored using an interferometric set-up which allows measurement for a limited range of strain values, although it was also reported that using an alternative approach to measuring the phase-shift based on the absolute position time-of-flight telemeter technique has been reported to be useful for strain measurement up to 15.8% strain in a single-mode PMMA fibre [30]. The high precision and immunity to fluctuation due to light source and bends in the fibre are advantages associated with interferometric-based techniques and since single-mode fibres are smaller in diameters compared to their multi-mode counterparts, they are less intrusive in cases where their embedment could lead to discontinuity in the geometrical build-up, for example, as embedded sensors in composite laminates. On the other hand, however, care and skill are required to successfully cleave and couple these single-mode fibres together, particularly in field environment, to ensure minimal coupling losses in addition to the inherent light loss due to the fibre material itself (typically 150 dB/m at 1500 nm).

An initial study by Silva-Lopez et al. [42] reported the sensitivity of dye doped single-mode PMMA fibres to strain and temperature using a Mach-Zehnder interferometric set-up. The study involved the loading of the fibre on a translation stage where the authors reported the phase sensitivity (1.31×10^7 rad/m) of the fibre for strain range of 0–0.04%. Kiesel et al. [30] has conducted further experiments using single-mode POF where the fibres were tested for their strain response to failure in order to determine the calibration coefficients at strain rates from 0.01/min to 3.05/min. The typical failure strains of the POF specimens used was 30%. They reported an upper limit of fringe visibility at 15.8% nominal strain in the fibres used indicating the maximum strain possible for the POF tested using the set-up in their study. The calibrated linear and nonlinear coefficients were found to be 1.37×10^7 rad/m and 3.1×10^6 rad/m, respectively.

Strain measurement based on the fibre stretching of 1 mm diameter multi-mode PMMA POF was demonstrated recently by Poisel [43]. The author monitored the phase shift of a sinusoidal signal in the fibre under various tensile loads (corresponding to increasing extension of the POF from 0 to 500 μm in steps of 50 μm). The interrogation set-up relies on detecting the difference in transit times through the polymer optical fibre (POF) using an electronic phase-shift detector. A resolution of 10 μm extension was reported to be possible under a tensile set-up. The simple system was also shown to be sensitive to bending loads and capable of measuring dynamic loading up to 5 Hz.

These studies have demonstrated the potential of interrogating a section of stretched fibre for strain monitoring. However, there is still limited work to demonstrate the higher dynamic strain measurement capability of the technique (which could be applied for SHM applications). In addition, interferometric sensing in general requires a stable platform due to their susceptibility to vibration-induced noise and hence further work will be required before this approach to POF strain sensing can be applied in real structures.

Other POF-Based Sensors for SHM Applications

Fibre Bragg grating (FBG) sensors are well known for their capability for strain measurement and the gratings are typically inscribed in silica optical fibres, the fundamentals of which are well documented elsewhere [47,48]. Briefly, it involves monitoring the shift in the peak or resonance wavelength of either the reflected or transmitted spectrum resulting from an applied strain on the fibre. The possibility of absolute strain measurement and multiplexing capability of grating-based optical fibre sensors has received considerable attention for application in structural health monitoring. For SHM applications, FBG has featured extensively in many published articles for monitoring a variety of physical parameters [1, 2]. Monitoring of strain in this class of fibre sensors relies on detecting the shift in the central wavelength of the reflection spectrum as a response to applied strain. The ability to induce a periodic refractive-index change in polymer-based optical fibres is a recent development and this has opened up further SHM applications using grating-based sensors since the availability of POF-based FBG sensors offers ease of handling, higher strain sensitivity (1.48 pm/$\mu\varepsilon$ compared to 1.15 pm/$\mu\varepsilon$ for silica-based FBGs at wavelength of 1523 nm) and higher strain limit (up to 3.61% strain compared to 0.5-0.6% strain for silica-based FBGs) [17–19, 45, 46]. Although more work and attention are required to further improve and understand the grating writing process in polymer fibres, their potential for SHM is evident. The possibility of using a POF-based FBG sensor for strain and temperature measurement have been reported and shown to allow strain monitoring up to seven times the measurement limit of its silica counterpart [45, 46]. The sensor used exhibited good reproducibility and reversibility over the large strain sensing range. In their studies, the authors showed that the polymer FBG sensor was able to sustain up to 3% strain before yielding and by using a combination of polymer and silica FBG, it was possible to discriminate between the effect of temperature and strain on the sensor read-out [46]. Using a matrix inversion technique and solving for change in temperature, $\Delta T = f \{\Delta \lambda_{POF-FBG}\}$ and change in strain, $\Delta \varepsilon = f \{\Delta \lambda_{Silica-FBG}\}$ which are formulated using the strain and temperature sensitivities of the POF-FBG and the silica-FBG, respectively, the technique allows the applied strain and temperature to be determined simultaneously.

As a response to the lack of commercial availability of suitable single-mode POF, Krebber et al. [44] demonstrated the use of a tapered multimode standard POF and non-tapered GI-POF for creating grating-based sensors. The authors successfully generated the gratings in the tapered POF (from original 0.75 mm fibre to 0.2 mm) and non-tapered GI-POF specimens which showed strong reflection spectrum at specific locations along the POF using an OTDR technique and reported that it was possible to measure the integral strain along the fibre and to resolve the local strain at various locations.

Writing long period gratings (LPGs) in POFs has been reported recently [19] although there appears to be little published work in the literature on their application for strain sensing. Li et al. reported success of inducing gratings in a highly photosensitive POF core using traditional photo-etching technology together with a low-cost high pressure mercury lamp as the light source but no mechanical test was carried to assess the mechanical response of the sensor produced. The transmission spectrum of the written LPG demonstrated a loss of 3 dB at a peak wavelength of 1568 nm; despite PMMA having peak transmission in the visible spectrum, presumably to interface with optical telecommunications spectrometers and lightsources using with glass optical fibres. For Bragg gratings in the visible spectrum, the required periodicity of refractive index gratings is 183 to 216 nm; for long period gratings, the periodicity is 100 to 500 μm so it can be easily created by direct laser writing. One reason for the lack of published work is due to short supply of good quality commercially available singlemode POF with low loss and doped for photosensitivity, another is the lack of availability of single mode POF connectors since the fibre is produced in non-standard diameters.

One of the most recent developments in polymer fibre research which has attracted significant attention is the microstructured POF (mPOF). This type of POF is easy to manufacture with consistent quality, but is not commercially available at present, although it may be available for research groups who are keen to find applications for their mPOFs. Interestingly LPGs have also been fabricated on mPOFs and tested for their mechanical response [49]. mPOF is ideal for FBG and LPG as it is easy to handle, since the mPOF has a large diameter and yet is endlessly singlemoded along its length with low loss compared with single mode POF. In mPOF, the microscopic air channels that run along the length of the fibre defines the light guiding mechanism in contrast to the variations in the refractive index of the fibre material in conventional POF [21]. Two of the main advantages of using LPGs in mPOFs over standard PMMA fibres include the possibility of tuning the sensitivities of the loss features corresponding to the different measurands of interest by altering the hole geometry in mPOFs [21] and the possibility of introducing directional bend sensitivity [24]. LPG has been introduced in mPOFs by mechanical imprinting using a 15 cm long template (with period of 1 mm) placed upon a heated fibre and in their study, Large et al. [25] studied the viscoelastic properties of the mPOF and reported that shift in spectral features for strains up to 8% was possible although above 2% strain a non-linear response was reported to be evident. Below 2% strain, the change in peak wavelength to strain could be computed to be approximately 1.2 pm/$\mu\varepsilon$. The study also showed a non-reversible deformation response due to strain-related creep following a 10 hour constant strain at 3% strain. In addition, the authors also reported a wavelength shift of up to 4 nm due to material relaxation at strains of 6%. In cases where intermittent straining was applied on the mPOF LPG sensor for up to 2% strain, the authors argued that the viscoelastic effects (time, strain rate, and strain magnitude) were small although time-dependent effects such as relaxation during constant strain and strain recovery could become significant. In SHM applications where these limitations are properly understood and deemed acceptable, mPOF LPG sensors may be successfully applied.

Conventionally, fibre Bragg gratings and long-period gratings sensors are interrogated using an optical spectrum analyser to detect the shift in wavelength of the reflected spectrum corresponding to the applied strain and/or temperature. However, the associated interrogation equipment is generally costly, involving the use of optical spectrum analysers and narrow band light sources such as laser diode. In an effort to circumvent the cost barrier to the wider adoption of grating-based sensors, Hwang et al. [49] recently demonstrated an interesting intensity-based set-up which utilizes two long-period fibre gratings and a core mode blocker between the two gratings. Although the fibers used were not polymer-based, this work is highlighted in view of the potential of the technique to be applied for interrogation of POF-based grating sensors. In principle, the core mode blocker acts as a band-pass filter to block the uncoupled light while the light that satisfy the phase-matching condition of the first LPG will be coupled to the cladding mode. The light in the cladding layer would then be effectively coupled back into the core by the second LPG. Strain is then applied on one of the LPGs. The relative change in the resonance wavelengths (change in the degree of spectrum overlap) due to the strain applied on one of the LPGs will result in either increase or decrease in the transmitted optical power. The power is monitored and then calibrated against the applied strain. In their work, the authors were able to show that the intensity transmitted through the fibre increases linearly with strain although the strain sensitivity (1.0×10^{-4} dB/$\mu\varepsilon$) was noted to be three time less than conventional single long-period grating fibre sensor [50]. The authors also noted small fluctuation in the transmitted power resulting from the unstable LED source though it was unclear whether the observed fluctuation affected the accuracy of the measurements significantly. Despite the lower sensitivity of the system (thought to be due to the broader bandwidth of the LPG relative to a laser diode), small fluctuation in transmitted power, the authors argued that the cost-effectiveness of the proposed sensor system would see its application for interrogation of grating-based sensors. With the possibility of introducing gratings in POFs as outlined by the reports above, it would be interesting to see further developments and field applications in intensity-based POF grating sensors for SHM applications using the technique proposed.

SUMMARY AND CONCLUSIONS

An overview of a selection of articles on the recent progress of the applications of plastic-based optical fibre sensors has been given. A tabulated summary highlighting the major works and a brief comparison of the types of sensors reviewed is shown in Table 1. The overview began with a brief introduction of the technology and sensing techniques used in POF sensors highlighting the various strengths and limitations associated with the different sensing schemes adopted. In view of the ease of handling, low cost and large strain capability of POF, significant interest in their use for structural health monitoring, specifically for measurement of strain, curvature, load, displacement, vibration and crack detection have been demonstrated by various workers. A number of different approaches in sensing interrogation such as intensity-based, interferometric-based, time-of-flight (OTDR-based) and gratings (wavelength-based) have been shown to be feasible for monitoring engineering materials such as concrete, masonry, fibre composites, geotextiles, metals, wood and plastics. There remains much work to be done to fully characterize the POF response to the various physical measurands encountered in SHM applications particularly with the recent development of new POF materials, types and interrogation techniques.

Table 1: Sensor performance and other comparison criteria

POF sensing technology	Type of POF used	Sensor type	Strain resolution	Dynamic range	Relative sensor cost	Relative equipment cost	Key features & remarks
Intensiometric							
[3, 4, 7, 9–11, 14, 28, 34]	SI MMF	Intrinsic	~50 µε	~1.2%	Very low (<US$1/m)	Very low (<US$200)	Simple design; high cost-effectiveness; demonstrated for flexural and axial loading, crack detection in concrete; monitoring low-velocity impact in advanced composites; POF embedded in fibre composites for axial strain measurement; capable of monitoring quasistatic and dynamic loading (~30 Hz or higher).
[12–15, 35, 36]	SI MMF	Extrinsic	~5 µε (liquid-filled type);	~0.12%	Very low (<US$1/m)	Very low (<US$200)	Simple design; high cost-effectiveness; though strain limit demonstrated up to 40%, higher strains possible; demonstrated for crack detection, monitoring crack-width opening, impact damage, geotextile strain, capable of monitoring quasistatic and dynamic loading (~1 kHz or higher).
			~100 µε (air-filled type)	~40% or more			
OTDR-based							

Ref	Fibre	Type	Resolution	Range	Cost/sensor	Cost of interrogation	Comments
[37–39]	SI MMF	Intrinsic	~0.5% Spatial resolution: ~2 m for 50 m POF to 5 m for 100 m POF	~50%–100%	Very low (<US$1/m)	Very high (>US$50,000)	Distributed measurement; High strain measurement, however, POF axial strainelastic limit ~5%; general ability to discriminate loading types based on backscatter shape trace; quasistatic measurement only. Demonstrated to be capable of supporting 100% strain applied at 5 locations along a 100 m POF.
[32, 33, 40, 41, 44]	SI & GI MMF	Intrinsic	~0.1% (SI POF) ~2% (GI POF) Spatial resolution: ~0.1–1 m for 100 m (GI POF)	~40%	Very low (<US$1/m)	Very high (>US$50,000)	Distributed measurement of 100 m (SI POF) and 500 m (GI POF); high strain measurement; demonstrated for monitoring geotextile strain, geotextile-retrofitted masonry structures; generally only quasistatic measurement only or very low frequency oscillation (~0.25 Hz) if high-speed OTDR unit used.
Interferometric							
[30, 31]	SM POF	Intrinsic	0.05 µε (for gauge length of 0.1 m)	15.8%	High (~USD$50/m)	High (~USD$6000–USD$7000)	Large strain measuring capability; High measurement precision; less intrusive due to smaller fibre diameter; Difficulty in cleaving and coupling the single mode fibres in the field; strain rate of ~3% strain/min possible.
[43]	SI MMF	Intrinsic	10 µε	500 µε	Very low (<US$1/m)	Low (<US$1000)	Simple concept; demonstrated dynamic monitoring on a rotor blade up to 5 Hz.

Category	Ref	Fibre	Type	Resolution	Range/Error	Availability	Interrogation	Remarks
Grating-based	[17, 45, 46]	PMMA; SM POF	Intrinsic	<1 με	3.6%	Currently not commercially available. Estimated cost: High (~US$100)	High to very high ~US$10,000–US$30,000 for OSA-based interrogators and broadband light source	Bragg gratings inscribed on in-house made dye-doped PMMA perform with UV-beam; strain measurement conducted by means of simple tension of POF. Scheme proposed allow temperature and strain to be measured simultaneously; depending on interrogator used, the sensor could monitor quasistatic and dynamic loading conditions.
	[20–25]	mPOF	Intrinsic	~1 με	8%	Currently not commercially available. Estimated cost: High >US$100	High to very high ~US$10,000–US$30,000 for OSA-based interrogators and broadband light source	Long period grating mechanically imprinted onto microstructured POF; linear response limited to 2% strain; creeping observed beyond 3%; viscoelastic effects noted above 2% but minimal below 2%; significant material relaxation under constant strain.
	[44]	SI MMF; GI MMG;	Intrinsic	Not reported	Not reported	Currently not commercially available.	High to very high ~US$10,000–US$50,000 for spectrometer, OTDR and broadband light source	Grating photo-induced onto tapered 0.2 mm SI-MMF and non-tapered GI POF with UV beam; OTDR used as signal interrogator instead of conventional OSA.
		mPOF	Intrinsic	Not reported estimated: ~0.02%	~1% for initial test	Estimated cost: High >US$100		Long period grating on mPOF integrated to geotextile; length of sensor extended using silica fibres.

ACKNOWLEDGMENTS

Funding by A*STAR-MPA through the National University of Singapore CORE centre under joint Grants no. R-264-000-226-490 and no. R-264-000-226-305 is acknowledged.

REFERENCES

1. K. S. C. Kuang and W. J. Cantwell, "The use of conventional optical fibres and fibre Bragg gratings for damage detection in advanced composite structures—a review," Applied Mechanics Reviews, vol. 56, pp. 493–513, 2003.
2. G. Zhou and L. M. Sim, "Damage detection and assessment in fibre-reinforced composite structures with embedded fibre optic sensors-review," Smart Materials and Structures, vol. 11, no. 6, pp. 925–939, 2002·
3. K. S. C. Kuang, W. J. Cantwell, and P. J. Scully, "An evaluation of a novel plastic optical fibre sensor for axial strain and bend measurements," Measurement Science and Technology, vol. 13, no. 10, pp. 1523–1534, 2002·
4. A. Babchenko, Z. Weinberger, N. Itzkovich, and J. Maryles, "Plastic optical fibre with structural imperfections as a displacement sensor," Measurement Science and Technology, vol. 17, no. 5, pp. 1157–1161, 200
5. N. Takeda, T. Kosaka, and T. Ichiyama, "Detection of transverse cracks by embedded plastic optical fiber in FRP laminates," in Smart Structures and Materials 1999: Sensory Phenomena and Measurement, vol. 3670 of Proceedings of SPIE, pp. 248–255, 1999.
6. N. Takeda, "Characterization of microscopic damage in composite laminates and real-time monitoring by embedded optical fiber sensors," International Journal of Fatigue, vol. 24, no. 2–4, pp. 281–289, 2002.·

7. A. Djordjevich, "Curvature gauge as torsional and axial load sensor," Sensors and Actuators A, vol. 64, no. 3, pp. 219–224, 1998.
8. G. Perrone, D. Perla, R. Gaudino, and S. Abrate, "Development of low cost intensity modulated POF based sensors and application to the monitoring of civil structures," in Proceedings of the 13th Plastic Optical Fibre Conference, pp. 444–449, 2004.
9. K. S. C. Kuang and W. J. Cantwell, "The use of plastic optical fibres and shape memory alloys for damage assessment and damping control in composite materials," Measurement Science and Technology, vol. 14, no. 8, pp. 1305–1313, 2003.
10. K. S. C. Kuang and W. J. Cantwell, "The use of plastic optical fibre sensors for monitoring the dynamic response of fibre composite beams," Measurement Science and Technology, vol. 14, no. 6, pp. 736–745, 2003.
11. K. S. C. Kuang, Akmaluddin, W. J. Cantwell, and C. Thomas, "Crack detection and vertical deflection monitoring in concrete beams using plastic optical fibre sensors," Measurement Science and Technology, vol. 14, no. 2, pp. 205–216, 2003.
12. K. S. C. Kuang, M. Maalej, and S. T. Quek, "An application of a plastic optical fiber sensor and genetic algorithm for structural health monitors," Journal of Intelligent Material Systems and Structures, vol. 17, no. 5, pp. 361–379, 2006.
13. K. S. C. Kuang, S. T. Quek, and M. Maalej, "Detection of impact induce damage in composite beams using plastic optical fibre sensors," in Proceedings of the 17th KKCNN Symposium on Civil Engineering, pp. 103–110, 2004.
14. K. S. C. Kuang, S. T. Quek, and M. Maalej, "Polymer-based optical fiber sensors for health monitoring of engineering structures," in Smart Structures and Materials 2005: Sensors and Smart Structures, vol. 5765 of Proceedings of SPIE, pp. 656–667, and 2005.
15. K. S. C. Kuang, S. T. Quek, and M. Maalej, "Assessment of an extrinsic polymer-based optical fibre sensor for structural

health monitoring," Measurement Science and Technology, vol. 15, no. 10, pp. 2133–2141, 2004.
16. K. S. C. Kuang, S. T. Quek, and W. J. Cantwell, "Use of polymer-based sensors for monitoring the static and dynamic response of a cantilever composite beam," Journal of Materials Science, vol. 39, no. 11, pp. 3839–3843, 2004.·
17. P. J. Scully, D. Jones, and D. A. Jaroszynski, "Femtosecond laser irradiation of polymethylmethacrylate for refractive index gratings," Journal of Optics A, vol. 5, no. 4, pp. S92–S96, 2003.
18. G. D. Peng and P. L. Chu, "Polymer optical fiber sensing," in Optical Information Processing Technology, vol. 4929 of Proceedings of SPIE, pp. 303–311, 2002.
19. Z. Li, H. Y. Tam, L. Xu, and Q. Zhang, "Fabrication of long-period gratings in poly (methyl methacrylate-co-methyl vinyl ketone-co-benzyl methacrylate)-core polymer optical fiber by use of a mercury lamp," Optics Letters, vol. 30, no. 10, pp. 1117–1119, 2005.
20. M. C. J. Large, L. Poladian, G. W. Barton, and M. A. van Eijkelenborg, Microstructured Polymer Optical Fibres, Springer, New York, NY, USA, 2007.
21. M. A. van Eijkelenborg, M. C. J. Large, A. Argyros, et al., "Microstructured polymer optical fiber," Optics Express, vol. 9, pp. 319–327, 2001.
22. H. Dobb, D. J. Webb, K. Kalli, A. Argyros, M. C. J. Large, and M. A. van Eijkelenborg, "Continuous wave ultraviolet light-induced fibre Bragg gratings in few- and single-moded microstructured polymer optical fibres," Optical Letters, vol. 30, pp. 3296–3298, 2005.
23. M. P. Hiscocks, M. A. van Eijkelenborg, A. Argyros, and M. C. J. Large, "Stable imprinting of long-period gratings in microstructured polymer optical fibre," Optics Express, vol. 14, no. 11, pp. 4644–4649, 2006.
24. H. Dobb, K. Kalli, and D. J. Webb, "Measured sensitivity of arc-induced long-period grating sensors in photonic crystal

fibre," Optics Communications, vol. 260, no. 1, pp. 184–191, 2006.

25. M. C. J. Large, J. Moran, and L. Yet, "The role of viscoelastic properties in strain testing using microstructured polymer optical fibres (mPOF)," Measurement Science and Technology, vol. 20, no. 3, Article ID 034014, 6 pages, 2009.

26. K. Murphy, S. Zhang, and V. M. Karbhari, "Effect of concrete based alkaline solutions on short term response of composites," in Proceedings of the 44th International Society for the Advancement of Material and Process Engineering Symposium and Exhibition, vol. 44, pp. 2222–2230, 1999.

27. D. Kalymnios, P. J. Scully, J. Zubia, and H. Poisel, "POF sensor overview," in Proceedings of the 13th International Plastic Optical Fibres Conference, pp. 237–244, 2004.

28. Y. M. Wong, P. J. Scully, R. J. Bartlett, K. S. C. Kuang, and W. J. Cantwell, "Plastic optical fibre sensors for environmental monitoring: biofouling and strain applications," Strain, vol. 39, no. 3, pp. 115–119, 2003.

29. R. J. Bartlett, R. Philip-Chandy, P. Eldridge, D. F. Merchant, R. Morgan, and P. J. Scully, "Plastic optical fiber sensors and devices," Transactions of the Institute of Measurement and Control, vol. 22, pp. 431–457, 2000.

30. S. Kiesel, K. Peters, T. Hassan, and M. Kowalsky, "Calibration of a single-mode polymer optical fiber large-strain sensor," Measurement Science and Technology, vol. 20, no. 3, Article ID 034016, 7 pages, 2009.

31. S. Kiesel, K. Peters, T. Hassan, and M. Kowalsky, "Behaviour of intrinsic polymer optical fibre sensor for large-strain applications," Measurement Science and Technology, vol. 18, no. 10, pp. 3144–3154, 2007.

32. S. Liehr, P. Lenke, K. Krebber, et al., "Distributed strain measurement with polymer optical fibers integrated into multifunctional geotextiles," in Optical Sensors 2008, vol. 7003 of Proceedings of SPIE, April 2008, article no. 700302.

33. P. Lenke, S. Liehr, K. Krebber, F. Weigand, and E. Thiele, "Distributed strain measurement with polymer optical fiber integrated in technical textiles using the optical time domain reflectometry technique," in Proceedings of the 16th International Conference on Plastic Optical Fibre, pp. 21–24, 2007.
34. P. Cortes, W. J. Cantwell, K. S. C. Kuang, and S. T. Quek, "The morphing properties of a smart fibre metal laminate," Polymer Composites, vol. 29, pp. 1263–1268, 2008.
35. K. S. C. Kuang, S. T. Quek, and W. J. Cantwell, "Morphing and control of a smart fibre metal laminate utilizing plastic optical fibre sensor and Ni-Ti sheet," in Proceedings of the 17th International Conference on Composite Materials, 2009.
36. K. S. C. Kuang, S. T. Quek, C. Y. Tan, and S. H. Chew, "Plastic optical fiber sensors for measurement of large strain in geotextile materials," Advanced Materials Research, vol. 47–50, pp. 1233–1236, 2008.
37. I. R. Husdi, K. Nakamura, and S. Ueha, "Sensing characteristics of plastic optical fibres measured by optical time-domain reflectometry," Measurement Science and Technology, vol. 15, no. 8, pp. 1553–1559, 2004.
38. K. Nakamura, I. R. Husdi, and S. Ueha, "Memory effect of POF distributed strain sensor," in Second European Workshop on Optical Fibre Sensors, vol. 5502 of Proceedings of SPIE, pp. 144–147, 2004.
39. T. Fukumoto, K. Nakamura, and S. Ueha, "A POF-based distributed strain sensor for detecting deformation of wooden structures," in 19th International Conference on Optical Fibre Sensors, vol. 7004 of Proceedings of SPIE, 2008, article no. 700469.
40. K. Krebber, P. Lenke, S. Liehr, J. Witt, and M. Schukar, "Smart technical textiles with integrated POF sensors," in Smart Sensor Phenomena, Technology, Networks, and Systems 2008, vol. 6933 of Proceedings of SPIE, 2008, article no. 69330V.

41. P. Lenke, S. Liehr, K. Krebber, F. Weigand, and E. Thiele, "Distributed strain measurement with polymer optical fiber integrated in technical textiles using the optical time domain reflectometry technique," in Proceedings of the 16th International Conference on Plastic Optical Fibre, pp. 21–24, 2007.
42. M. Silva-Lopez, A. Fender, W. N. MacPherson, et al., "Strain and temperature sensitivity of a singlemode polymer optical fibre," Optics Letters, vol. 30, pp. 3129–3131, 2005.
43. H. Poisel, "POF strain sensor using phase measurement techniques," in Smart Sensor Phenomena, Technology, Networks, and Systems 2008, vol. 6933 of Proceedings of SPIE, 2008, article no. 69330Y.
44. K. Krebber, P. Lenke, S. Liehr, et al., "Technology and applications of smart technical textiles based on POF," in Proceedings of the 17th International Conference on Plastic Optical Fibre, 2008.
45. H. Y. Liu, H. B. Liu, and G. D. Peng, "Strain sensing characterization of polymer optical fibre Bragg gratings," in 17th International Conference on Optical Fibre Sensors, vol. 5855 of Proceedings of SPIE, pp. 663–666, 2005.
46. H. B. Liu, H. Y. Liu, G. D. Peng, and P. L. Chu, "Strain and temperature sensor using a combination of polymer and silica fibre Bragg gratings," Optics Communications, vol. 219, no. 1–6, pp. 139–142, 2003.
47. K. O. Hill, Y. Fuji, D. C. Johnson, and B. S. Kawasaki, "Photosensitivity in optical fiber waveguides, application to reflection filter fabrication," Applied Physics Letters, vol. 32, pp. 647–649, 1978.
48. G. Meltz, G. G. Morey, and W. H. Glenn, "Formation of Bragg grating in optical fibers by a transverse holographic method," Optics Letters, vol. 14, pp. 823–825, 1989.
49. D. Hwang, L. V. Nguyen, D. S. Moon, and Y. J. Chung, "Intensity-based optical fiber strain sensor,"Measurement

Science and Technology, vol. 20, Article ID 034020, 6 pages, 2009.

50. M. de Vries, V. Bhatia, T. D'Alberto, V. Arya, and R. O. Claus, "Photoinduced grating-based optical fiber sensors for structural analysis and control," Engineering Structures, vol. 20, no. 3, pp. 205–210, 1998.

Chapter 2

Investigation of Wave-Structure Interaction Using State of the Art CFD Techniques

Jan Westphalen[1], Deborah M. Greaves[1], Alison Raby[1], Zheng Zheng Hu[1], Derek M. Causon[2], Clive G. Mingham[2], Pourya Omidvar[3], Peter K. Stansby[3], and Benedict D. Rogers[3]

[1]School of Marine Science and Engineering, Marine Institute, University of Plymouth, Plymouth, UK

[2]School of Computing, Mathematics and Digital Technology, Manchester Metropolitan University, Manchester, UK

[3]School of Mechanical, Aerospace and Civil Engineering, University of Manchester, Manchester, UK

ABSTRACT

The suitability of computational fluid dynamics (CFD) for marine renewable energy research and development and in particular for simulating extreme wave interaction with a wave energy converter (WEC) is considered. Fully nonlinear time domain CFD is often considered to be an expensive and computationally intensive option for marine hydrodynamics and frequency-based methods are traditionally preferred by the industry. However, CFD models capture more of the physics of wave-structure interaction, and whereas traditional frequency domain approaches are restricted to linear motions, fully nonlinear CFD can simulate wave breaking and overtopping.

Furthermore, with continuing advances in computing power and speed and the development of new algorithms for CFD, it is becoming a more popular option for design applications in the marine environment. In this work, different CFD approaches of increasing novelty are assessed: two commercial CFD packages incorporating recent advances in high resolution free surface flow simulation; a finite volume based Euler equation model with a shock capturing technique for the free surface; and meshless Smoothed Particle Hydrodynamics (SPH) method.

These different approaches to fully nonlinear time domain simulation of free surface flow and wave structure interaction are applied to test cases of increasing complexity and the results compared with experimental data. Results are presented for regular wave interaction with a fixed horizontal cylinder, wave generation by a cone in driven vertical motion at the free surface and extreme wave interaction with a bobbing float (The Manchester Bobber WEC). The numerical results generally show good agreement with the physical experiments and simulate the wave-structure interaction and wave loading satisfactorily. The grid-based methods are shown to be generally less able than the meshless SPH to capture jet formation at the face of the cone, the resolution of the jet being grid dependent.

INTRODUCTION

In the design of floating offshore WEC structures, many of the same engineering issues arise as for floating offshore structures used in the oil and gas industry. There are some similarities between the design challenges in the two industries and some of the techniques and tools developed within the oil and gas sector can be applied to wave energy. However, there are also significant differences. WECs are usually designed to have moving parts that react to wave motion; WECs are generally much smaller than offshore structures; the water depth is usually shallower and WECs are generally expected to be deployed in arrays, so the design is not a "one off". The structural design of offshore floating wave energy converters (WECs) generally involves three main and interacting parts of the device: the floating absorber, the mooring arrangement and the power-take-off (PTO) with its associated control system.

Unlike large offshore structures in which motions and forces are minimised, WECs are designed to perform large motions in small and intermediate seas for efficient power conversion, but to survive extreme waves in the event of a storm. These design parameters require mooring arrangements that hold the device in place but do not restrict the motion of the WEC in its energy capturing mode. The dynamics of the moorings, including the motion of the cables, risers, weights and floats, may be highly non-linear in terms of hydrodynamic and structural interaction, and this can affect the response of the device negatively (or positively). The PTO system, however, directly affects the dynamics of the WEC as it converts energy.

In the engineering design of WECs, the three areas of device hydrodynamics, mooring and PTO/control strategy can affect each other considerably and therefore should be considered together. The hydrodynamics can be described using the equations of motion to represent, for example, a single-body heaving device as a mass-spring-damper system, in which the mooring forces are included in the system stiffness and the PTO forces as external forces. Added mass and damping are shape and frequency dependent and can be evaluated using standard naval architecture techniques by physical tank tests, or by numerical methods, which solve the initial boundary value problem for radiation and diffraction for a given geometry [1] [2] . The assumptions for these numerical methods to be valid are: irrotational, inviscid fluid flow, linear wave theory and small body motion. By construct, in the frequency domain the waves are monochromatic and sinusoidal, which limits their applicability for a WEC in real sea conditions.

Modelling in the time domain is necessary for WEC control studies and for investigation of nonlinear wave interaction. Cummins [3] converted the frequency domain coefficients in the time domain by calculating impulse response functions (IRF). The IRF describes the response of the system after an initial impulse and can be used in the time-domain formulation of the equations of motion. With this approach it is possible to simulate polychromatic seas. However, the same assumptions of linear theory and irrotational, inviscid fluid flow apply. When several waves are superposed, these assumptions can be violated, as waves may become large, and especially when the wave frequency is close to the natural frequency of the WEC, unrealistically large body motions may occur because friction and viscous losses are not usually taken into account.

Here, we focus on the hydrodynamics of the WEC body and its interaction with an extreme wave and investigate a hierarchy of CFD approaches applied to this case. The design of the WEC with regards to its survivability demands a nonlinear description of the hydrodynamics. Here, steep and large waves that can generate significant water impact on the structure need to be modelled. This may involve wave breaking and green water on the deck of the device. Viscous effects may become large and also the body motions may become highly non-linear, as a floating device can become fully submerged or airborne.

As argued above, when investigating the hydrodynamics of extreme wave-structure interaction, linear theory is of uncertain accuracy and fully non-linear numerical methods are required. Mathematically, such problems can be described by the Navier-

Stokes equations incorporating a free surface. These include viscous effects and, depending on the discretisation of the equations, surface effects such as wave breaking can be simulated. When viscous effects are judged to be not important and are neglected, the Navier-Stokes equations reduce to the Euler equations. Both sets of equations can be discretised on a two or three-dimensional mesh and numerical techniques such as the well-known Finite Volume (FV) or Finite Element (FE) formulations can be used to compute the fluid velocities and pressures at every mesh cell or node. Bodies can be modelled as cavities in the mesh and body motion can be simulated by dynamic body fitted meshes. Special attention needs to be taken when modelling the free surface motions, and Lagrangian surface tracking or Eulerian surface capturing may be used. The volume of fluid (VoF) surface capturing method is often used in CFD and involves the introduction of an additional variable, the volume fraction, which is advected with the flow and represents the fraction of water in a given cell. At the free surface, cells are partially filled and the volume fraction lies between 0 and 1. This approach has been used successfully to simulate the motions of a ship in waves, as reported by Hadzic et al. [4] . The alternatives to the Eulerian grid-based methods are Lagrangian methods such as Smoothed Particle hydrodynamics (SPH); these are meshless methods and do not require special treatment of a free surface.

In this paper we use three Eulerian based methods, two that solve the Navier-Stokes equations and one the Euler equations on a three dimensional mesh, and Lagrangian SPH methods. All of the techniques are used to simulate fluid structure interaction problems of increasing complexity. First, a fixed horizontal cylinder in regular waves is modelled. The time histories of the vertical forces are compared to physical experiments and potential theory [5] . The next experiment is a cone in vertical motion that radiates waves at the still water surface. It follows a prescribed displacement of the form of a Gaussian wave packet [6] .

The last set of results compares the motions of a freely floating body interacting with extreme large waves and a counterweight, which is connected to the main float by a pulley-rope system. This case shows the capabilities of the three Eulerian CFD methods to simulate a floating body in one and two degrees of freedom in very non-linear waves, but also the interaction with a counterweight. This counterweight is represented by a force acting on the body, which can provide an option to also include non-linear PTO and mooring forces, for example to test controllers or evaluate non-linear hydrodynamic interaction between the main device and its auxiliaries.

Section 2 describes the three CFD approaches used here, namely the Navier-Stokes/Finite Volume, the Euler/ Cartesian-Cut-Cell and SPH solvers. The test cases including the results are described in Section 3 and conclusions are discussed in Section 4.

NUMERICAL MODELS

Finite Volume (FV) Method

The Finite Volume solver [7] is indicated by the acronym CV and refers to the commercial code STAR-CCM+. It uses the Navier-Stokes equations discretised on a 3-dimensional mesh to calculate the velocities and pressures in the flow field in a segregated iterative way. The domain, here a numerical wave tank (NWT) including the structure, such as a horizontal cylinder, is subdivided into discrete volumes. The surface and volume integrals performed on the control volumes are used to calculate the solution variable values at the centre node of each CV as illustrated in Figure 1 [8] The variables are stored in a collocated manner, such that the pressures and velocities are computed at the same locations. To avoid the so-called "checker-board" effect, which might occur for non-staggered variable arrangements [9] , the pressures and velocities are calculated using a segregated iterative approach [10] .

Free surface calculations are performed using the VoF method [11] and an additional equation for advection of the volume fraction, c, is solved. For a CV that is fully filled with water, the volume fraction, c = 1.0; when empty of water, the volume fraction c = 0; for cells at the interface containing fractions of both fluids, the sum of both fractions is unity. To find the location of the water surface, which is taken as the iso-value of the water fraction c = 0.5, a high resolution interface capturing scheme (HRIC) is applied. This is dependent on the local Courant number of the cell and thereby on the time step. It is similar to the compressive interface capturing scheme for arbitrary meshes (CICSAM) [8] [12] -[14] .

Meshes comprising hexahedral cells are used for the simulations presented here. A typical mesh section can be seen in Figure 2(a). The cells are refined in steps of two, so that one coarse cell is split into two smaller cells occupying the same face length in two dimensions. Figure 2(b) shows the mesh close to the boundary of a cylinder, which is modelled as a cavity in the mesh. Prism layers are used to resolve the hydrodynamics in the vicinity of the surface properly. Where the boundary-fitted prism layers meet the rest of the mesh, arbitrary cell shapes occur, as the boundary-fitted mesh structure is effectively cut out of the initially hexahedral mesh.

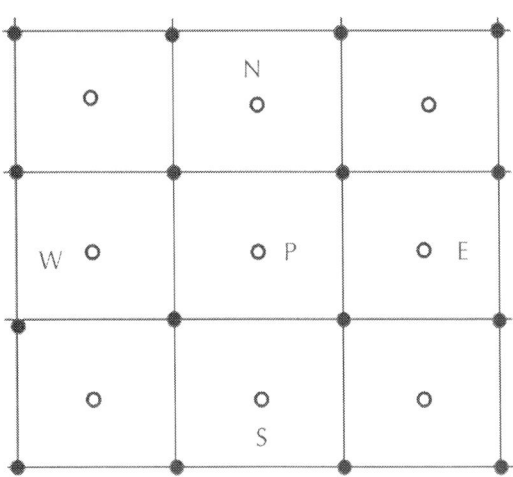

Figure 1: Arrangement of velocity components on a control volume element of a structured grid at the central node, element faces, and element vertices.

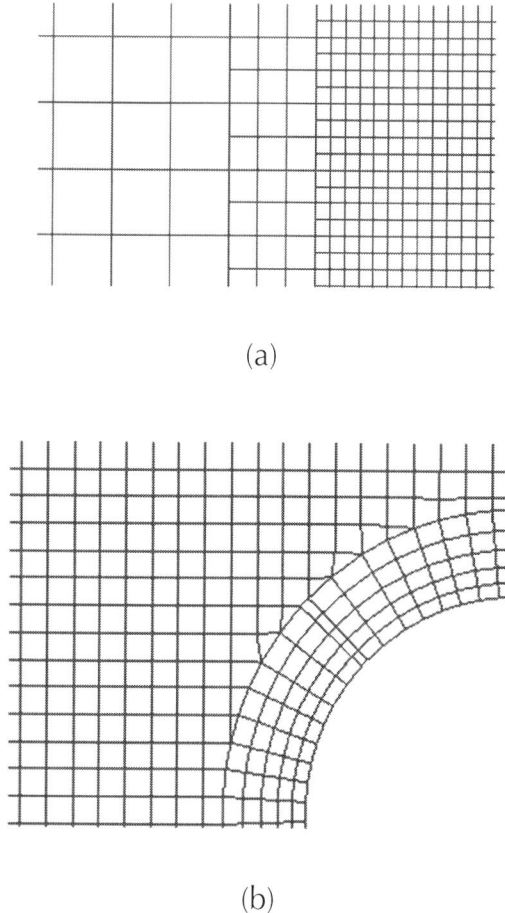

Figure 2: Mesh refinement (a) and prism layer close to a body, modelled as a cavity (b).

When mesh motion is included in the simulation, as for the single float in extreme waves described in Section 4, the whole domain is displaced according to the motion of the float. Thus, the connectivity of this mesh is fixed, the cut cells are not recomputed and re-meshing or dynamic meshing is not necessary at each time step. The mesh motion is taken into account by calculating the face velocities required in the discretisation of the governing equations as the difference between the fluid velocity and the face velocity resulting from the mesh motion.

Control-Volume Finite Element (CV-FE) Method

The Control-Volume Finite Element (CV-FE) approach [15] used here is the commercial package ANSYS CFX. It combines the Finite Volume method's control volume formulation with the Finite Element method by using shape functions and a vertex-centred discretisation scheme. The shape functions are used to describe the distribution of a variable across the CV [15] -[17] . As described for the Finite Volume method in the previous section, the CV-FE also solves the Navier-Stokes equations. The general transport is discretised using a collocated variable arrangement on a 3-dimensional grid containing hexahedral cells. The CVs are arranged around the mesh nodes (vertex-centred) as shown in Figure 3 and thereby this technique ensures the conservation of flow quantities such as mass and momentum.

As above, the volume fraction field is solved using the VoF formulation [10] . However, in the CV-FE approach, the fluid interface is calculated using a compressive interface capturing scheme [15] [18] -[20] , which is dependent on the filling level of the surrounding cells rather than the Courant number as is the case for CICSAM[12] . This means that the resolution of the fluid interface is time step independent and allows for larger time steps. The discretised equations are solved using a fully coupled system of linear equations, in which the pressure and velocity are linked to avoid decoupling and non-physical solutions that may otherwise occur [21] .

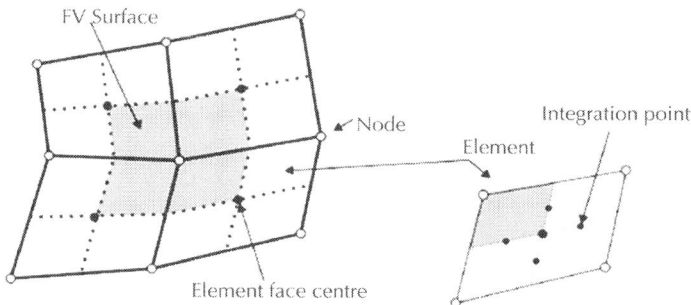

Figure 3: Vertex-centred discretisation of variables for Navier-Stokes/CV-FE solver.

In Figure 4(a), a typical mesh used for the CV-FE solver is shown, in which meshes with only hexahedral cells are used. In two dimensions, the cell refinement divides each cell into three, and creates transition layers such that each cell face coincides with one cell face on the refined side. A structure in the domain is represented using a body-fitted mesh of hexahedral cells, which may be distorted close to the structure boundary as seen in Figure 4(b). This may influence the initial solution of the free surface, if it cuts through these distorted cells. For cases involving body motion, the mesh is moved at each time step by moving the nodes and edges of the mesh according to the local velocity field, before the fluid dynamics equations are solved. In this way, the mesh connectivity is not changed, but the mesh cells are stretched.

Smoothed Particle Hydrodynamics

Smoothed Particle Hydrodynamics (SPH) is a flexible Lagrangian and meshless technique for CFD simulations that can be used in complex problems. In this method the fluid system is represented by a set of particles which have individual material properties and move according to governing conservation equations [22] . In the Lagrangian method, any flow variable is expressed as $\phi = (x,t)$, where x is the point vector of the particle at the reference time t =0.

The major advantage of using SPH is in dealing with large deformations and distorted free-surface problems. There is no mesh construction in SPH, therefore in certain problems, for instance simulation of waves, the SPH method may be easier to develop and use than Eulerian methods. Also, there is no need for special treatment of the free surface in order to simulate highly nonlinear and potentially violent flows, such as breaking waves. Furthermore, the equations used in SPH are quite simple in comparison with other methods. However, the computational cost is one of the disadvantages of SPH; the time step is much smaller than other methods due to using an explicit time integration scheme.

Following the work of [23] , the governing equations for compressible flow are solved in conservative form with an Arbitrary-Lagrangian Eulerian (ALE) SPH scheme. Here, a Riemann solver is used for each particle-particle interaction since pressure fields can be predicted satisfactorily and the classical SPH formulation may cause propagating waves to decay in the channel [24] . The interaction between each particle pair is solved as a 1-D Riemann problem. Herein, we use a parallel version of the SPHysics code [24] where this Riemann problem is solved using an HLLC approximate Riemann solver with MUSCL-based upwinding [25] and a -limiter [26] [27] .

The SPHysics code [28] has a choice of kernel function. Here, we use the cubic kernel function, which is found to be the best choice for propagation of waves in a channel [22] . The cubic kernel is known to exhibit tensile instability, and has to be corrected, according to [27] in order to avoid particle clumping. The symplectic algorithm [22] [28] , often known as kick-drift-kick, is used as the time stepping method. Here, kick is the change in velocity v by the force and drift is the change in coordinate, r.

The repulsive force boundary condition [28] [29] , uses boundary particles to exert forces on fluid particles according to a specified function acting normal to the wall. Boundary particles experience an equal and opposite force to the repulsive force that they exert on surrounding fluid particles. Based on the technique of Monaghan et al. [29] for simulation of rigid floating bodies, the force on each boundary particle is computed by summing up the contribution from all the surrounding water particles within its kernel. The boundary particles within the rigid body are then moved by integrating in time the equations of motion of the body in the translational and rotational degrees of freedom. It can be shown that this technique conserves both linear and angular momentum [22] [29] .

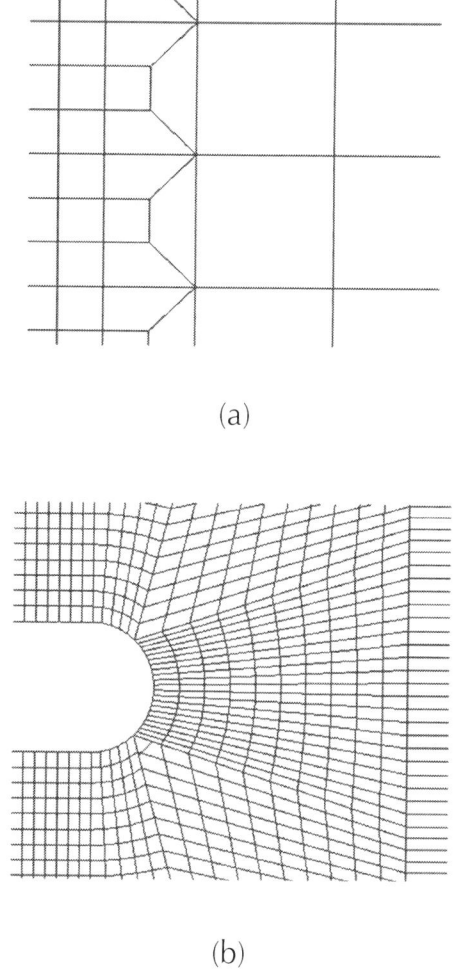

(a)

(b)

Figure 4: Mesh refinement (a) and body-fitted mesh close to the structure, modelled as a cavity (b).

Incompressible SPH (ISPH)

For virtually incompressible flows such as water the divergence free condition for continuity may be directly imposed by the projection method of Chorin [30], first applied by Cummins and Rudman[31] in SPH. This has the advantage of much reduced pressure noise, larger time steps, improved accuracy but the disadvantage of a Poisson solver for pressure. The present method is described by Lind et al. [32] with some improvements by Skillen et al. [33]. A second-order time marching scheme is applied, where both the density and mass of the incompressible particles are constant and the stabilised bi-conjugate gradient method is used to solve the linear system that results from the discretised form of the pressure Poisson equation.

There is a further difficulty associated with highly accurate ISPH; instability can occur due to particle clustering, e.g. near stagnation points. This may be avoided without loss of accuracy or convergence by particle shifting, equivalent to remeshing. This may be slight shifting across streamlines [34] or more generally by Fickian shifting along lines of decreasing particle density [32] which is suited to free surface flows.

Cartesian Cut Cell Method

The AMAZON-3D numerical wave tank (NWT) is based on the free surface capturing method for two fluid flows with moving bodies developed by Qian et al. [35], in which is demonstrated a rigid 2D wedge-shaped body entering and exiting calm water including total immersion. The AMAZON-3D code uses a Cartesian cut cell method to provide a boundary-fitted grid for both static and moving boundaries in 3D. The main advantages of the Cartesian cut cell approach have been outlined previously [36] [37], including particularly its flexibility for dealing with complex geometries and moving bodies. There is no requirement to re-mesh globally or even locally for a moving boundary problem which only requires changes locally at cells in the background Cartesian mesh that are cut by the moving boundary contour. The AMAZON-3D code is based on the integral form of the Euler equations for 3D incompressible flow with variable density. The free surface is treated as a contact surface in the density field that is captured automatically during a time-marched calculation without special provision in a manner analogous to shock capturing in compressible flow. A time-accurate artificial compressibility method and high resolution Godunov-type scheme replaces the pressure correction solver used in many current VoF methods. AMAZON-SC can handle break-up and recombination of the free surface as well as air entrainment into the water and, in principle, associated local compressibility effects. The total force is obtained by integration of the pressure field along the body given by

$$F = -\int_{S_b} pn dS$$

, where S_b is the body surface and is defined by the cut cell surface.

RESULTS

In order to assess the suitability of the four different CFD approaches to simulate WEC survivability design scenarios, they are applied to a series of benchmark test cases of increasing complexity. In each case, data from physical experiments are available and used for comparison. Grid convergence has been studied in comparisons of free surface wave simulations using FV and CV-FE, in which regular waves are simulated by using different meshes of hexahedral, polyhedral and tetrahedral shape with different resolution and is summarized in [38]].

Fixed Horizontal Cylinder in Regular Waves

Dixon et al. [5] carried out physical tank tests in order to improve Morison's formula for the calculation of the forces on a horizontal cylinder. For different levels of cylinder submergence and different wave amplitudes the vertical forces acting on the cylinder in waves were recorded. Here, the four CFD techniques are applied to a selection of these tests and compared with the experimental results.

The forces calculated using each of FV, CV-FE, SPH and AMAZON are compared for different wave amplitudes and cylinder axis depths. The properties including the position of the cylinder below still water level, d, the wave steepness, kA, the product of the wave number and water depth, kh and the Keulegan-Carpenter number K_c for each case are summarised in Table1 The relative amplitude A' and axis depth, d', are non-dimensionalised by the cylinder diameter D. To compare the numerical results with those obtained by Dixon et al. [5], the vertical forces, F_z, on the cylinder resulting from drag and pressure on the surface are derived. In the physical experiments, the forces were measured over one wave cycle. The force predicted by the CFD methods once steady-state was reached, F_z, is non-dimensionalised using the following expression

$$F' = \frac{F_z}{g\rho(1/4\pi D^2 l)}$$

(1)

with F_z being the measured vertical force on the cylinder, g, the acceleration due to gravity, ρ, the density of water, D, the cylinder diameter and l, the length of the cylinder.

As in the physical experiments, linear regular waves are generated in the NWT to interact with the structure. For the Eulerian grid-based approaches described in sections 2.1, 2.2 and 2.4, the wave velocity components u and w and the surface elevation η are described by

$$u = \frac{gAk\cosh(k(z+h))\cos(kx-\omega t)}{\omega\cosh(kh)}$$

(2)

$$w = -\frac{gAk\sinh(k(z+h))\sin(kx-\omega t)}{\omega\cosh(kh)}$$

(3)

and

$$\eta = A\cos(kx - \omega t) \tag{4}$$

and are applied to the water component at the upstream end of the 3-dimensional NWT using a velocity inlet boundary condition. The velocities for the air component are set to 0.0 m/s. The top boundary is a pressure outlet, the sides are modelled as symmetry planes and the bottom, the cylinder boundary and the downstream end of the domain are defined as walls. The total number of cells for the FV meshes is approximately 114,599 and the CV-FE meshes contain approximately 695,375 cells. The number of cells on the boundary of the cylinder itself, however, is 250 for the FV solver and 236 for CV-FE. The time step to achieve a converged solution is found to be 0.001 s and 0.005 s for FV and CV-FE respectively. Details of the NWT for each of the codes used are summarised in Table2 In AMAZON SC, the air part of the left-hand boundary and the top and right boundary are specified as non-reflecting boundary conditions allowing air to leave or enter the domain. ditions allowing air to leave or enter the domain. The cylinder surface is defined as a reflective wall boundary. The remaining sides are slip boundaries. The NWT contains approximately 458,850 cells with a minimum edge length of 0.015 m. A time step of 0.00025s is applied. For the SPH method the calculations are carried out in a 2-dimensional NWT using 7800 particles in the domain. The boundaries at the wave maker, the bottom, the downstream end and the cylinder are treated as walls and waves are generated by moving the upstream wall similar to a piston wave maker.

Table 1: Properties of horizontal cylinder cases

	1	2
d'	0	-0.3
A'	0.5	0.2
kA	0.2	0.01
kh	1.61	1.61
N_{KC}	3.1	1.3

Table 2: Details of the NWT used by each code for the horizontal cylinder cases

	CV_FE	AMAZON SC	SPH
Domain dimensions (m)	2.5 * 2.5 * 2.0	2.0 * 1.6 (2D)	4.0 * 4.0 * 1.0
No. of cells/particles	820,000 (3D)	32,000 (2D)	272,000 (3D)
Time step (s)	0.00005	0.00005	0.0001
Convergence error/criteria	0.0001	0.0001	N/A
Computer architecture used	HPCx cluster on 16 1.5 GHz processors	1 processor, 2.5 GHz Macbook Pro computer	1 processor of 1.2 GHz Linux workstation
CPU time: simulation time	3.5 days: 6 s	11 hours: 6 s	1.5 days to 6 s

In Figures 5 and 6 the time-histories of the non-dimensionalised vertical forces, F', over one wave period are shown for all solvers. For the mesh methods, both mesh and time step convergence are achieved for the results presented. The ISPH method is converged regarding the number of particles in the NWT and the time step. Convergence studies for the different CFD approaches are reported in the literature [35] [39] [40] . It was found by Westphalen et al. [39] that 14 cells in the vertical direction are sufficient to resolve the free surface accurately.

For case 1, d' = 0.0 and A' = 0.5, the three codes give very good agreement with the experimental data. The CV-FE results follow the experimental data well and the FV also agrees well, although a third peak is evident in the numerical result but not visible in the experiment data. The AMAZON SC result also represents the force characteristic reasonably well, although the second peak is not resolved well and occurs late and reduced in size, however, the second trough is generally well reproduced. The ISPH result gives a slightly noisy signal and the first trough is shallower and offset compared with the experiment, but the general trend is in reasonable agreement with the other data. For case 2, d' = −0.3 and A'= 0.2 the results look even better and each of the mesh-based solutions follows the experimental data well and reproduces the observed characteristics. As before, the ISPH results are noisier than the other techniques and this is would be reduced by increasing particle resolution but with increasing computational cost. Table 2 gives details of the CPU run time and computer architecture used by each of the codes.

Figure 7 shows snapshots of the surface elevation around the cylinder for case 1 (d' = 0.0) at different times calculated by the FV solver and the AMAZON code. The different codes show similar free surface behaviour at each time step, with overtopping of the cylinder t/T = 0.36. At time step, t/T = 0.6, air entrainment can be observed in the FV solution. Figure 8 shows the free surface profile in the tank and particle distributions calculated using the ISPH method at two different time instants for case 1. The solution in Figure 8(a) is at a time instant between those in Figures 7(i) and (j), and Figure 8(b) corresponds with Figure 7(l). The wave propagation near the cylinder has clearly been altered by the presence of the cylinder. Since the ISPH simulations are mono-phase (i.e. only water particles) the compressibility of air around the cylinder is not taken into account.

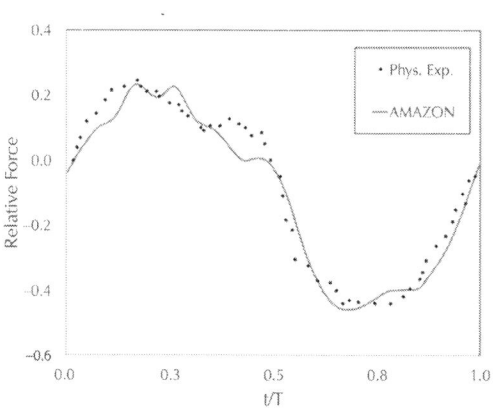

(a)

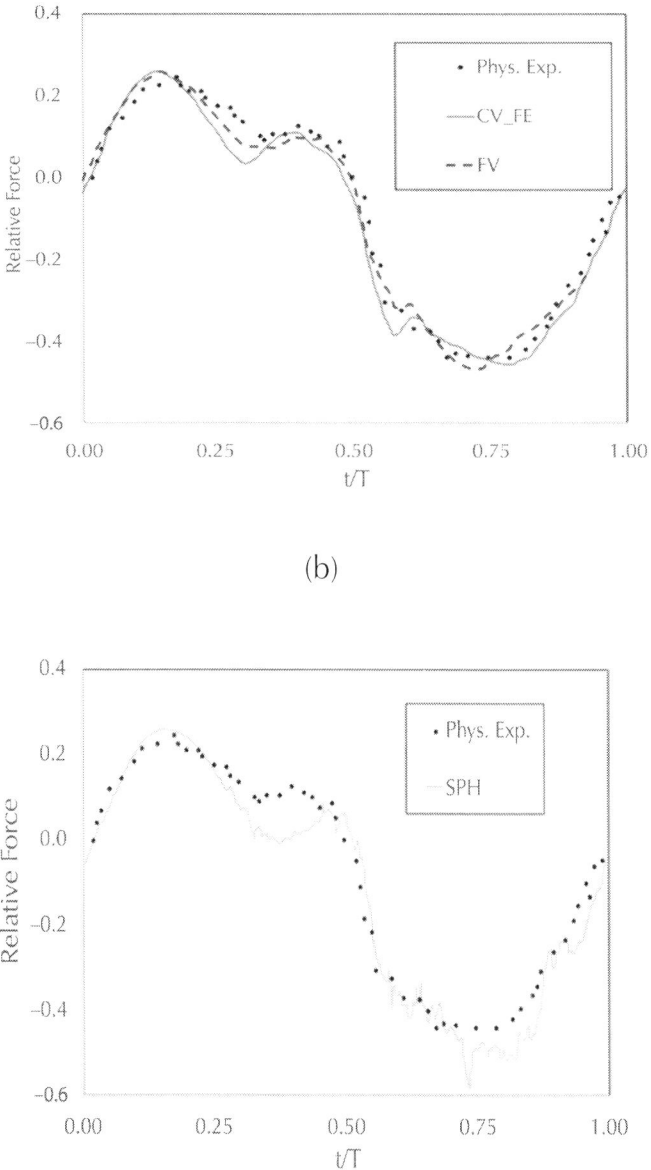

(b)

(c)

Figure 5: Relative vertical forces on horizontal cylinder, Case 1 (d' = 0.0, A' = 0.5); a) CV_FE and FV, b) AMAZON, c) SPH.

The cylinder is half submerged in case 1 and three quarters submerged in case 2 and the wave amplitude in case 1 is larger than in case 2, such that the wave overtops the cylinder in case 1 but not in case 2. Thus, it is to be expected that the results predicted by the numerical codes for the less challenging case 2 will be better than those for case 1 in which wetting and drying of the cylinder occurs, as seen in Figures 7 and 8.

Driven Motion: Oscillating Cone in Still Water

For the simulation of floating bodies, the eventual aim is to simulate the motion of floating WECs in extreme waves, and thus it is important to be able to calculate the forces on a moving body and the surface elevations around it correctly. For the second test case, a cone shaped body positioned with its vertex at the initially still water surface is chosen, for which physical tank tests are described by Drake et al. [6] . The motion of the cone is driven and not influenced by the forces generated on its surface by the surrounding fluids. In the experiments, the vertical forces on the cone surface due to its motion and the relative water surface elevation at a distance of 0.02 m from the cone surface are recorded. Here, comparisons between AMAZON, the CV-FE solver, SPH and the physical experiments are shown.

The motion of the cone is defined by the displacement d(t) from the initial position at t = 0 s following the form of a Gaussian wave packet, which is described by

$$d(t) = A \sum_{n=1}^{N} Z(\omega_n) \cos\left[\omega_n (t - t_0) - \frac{h\pi}{2}\right] \Delta\omega_n \quad (5)$$

where

$$Z(\omega_n) = \frac{1}{\frac{\omega_n}{2\pi}\sqrt{2\pi}} \exp\left[-\frac{(\omega_n - \omega_0)^2}{2\left(\frac{\omega_0}{2\pi}\right)^2}\right] \quad (6)$$

with h = 0 or 1. A denotes the largest excursion from the still water level. N is the number of frequency components and ω_n is the appropriate circular frequency. The central circular frequency ω_0 [rad/s] is defined by

$$\omega_0 = \frac{m\pi}{3} \quad (7)$$

with m being an integer between 1 and 12. m effectively controls the linearity of the case. The larger m is and thereby the central frequency, the more non-linear the dynamics become. The results presented here are for h = 1, m = 6 and 9 and A = 0.05 m.

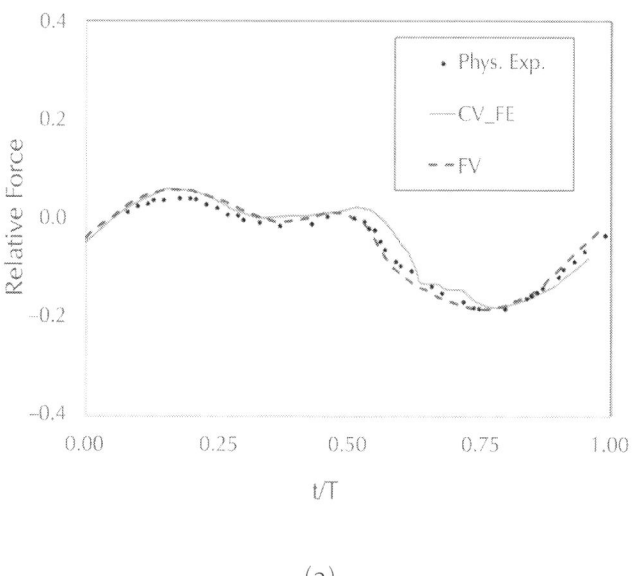

(a)

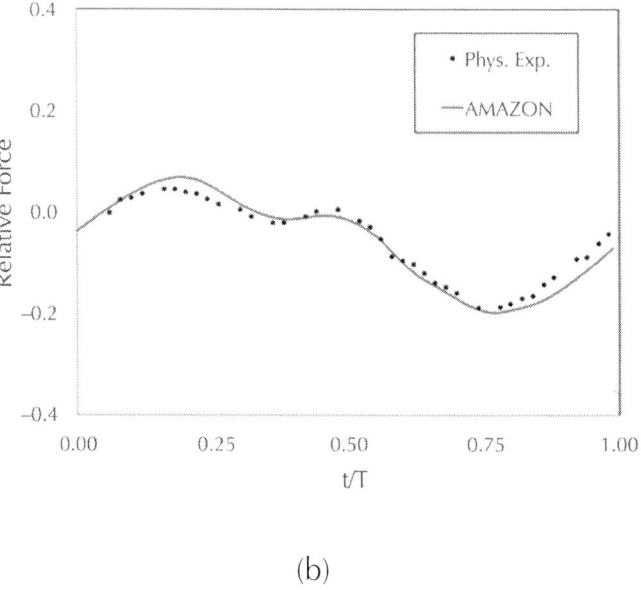

(b)

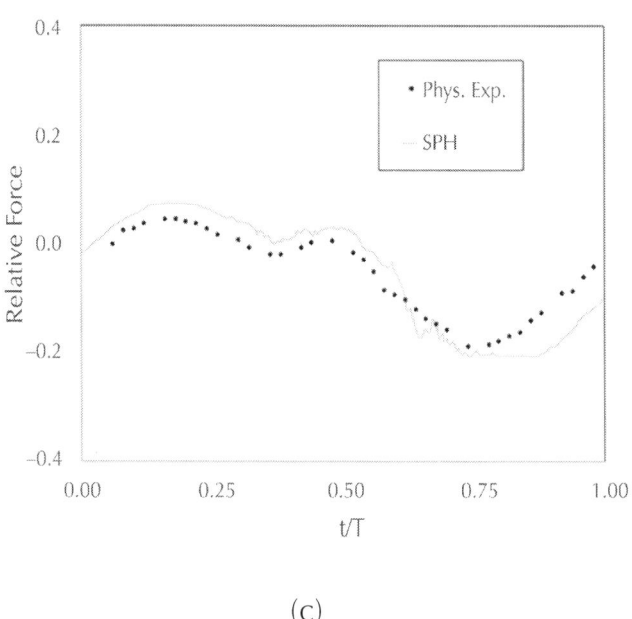

(c)

Figure 6: Relative vertical forces on horizontal cylinder, Case 2 (d' = −0.3, A'=0.2); a) CV_FE and FV, b) AMAZON, c) SPH.

For the CV-FE approach the simulations are performed in a three-dimensional domain with a length and width of 2.5 m and a height of 2.0 m. The cone is placed in the centre, as can be seen in Figure 9 and has a maximum diameter equal to 0.6 m and a dead rise angle of 45°. The slope itself is 0.3m high. The initial draught of the cone is 0.15 m at a water depth of 1.0 m. The cone is modelled as a cavity in the mesh. The outer boundaries, the bottom of the tank and the cone are modelled as free slip walls. The top boundary is defined as a pressure outlet with constant atmospheric pressure. The mesh consists of 820,000 hexahedral cells, where the regions around the water surface and the cone surface are highly refined to achieve cell edges of approximately 0.01 m. The outer regions are relatively coarse to save computational resources and encourage numerical damping, thus avoiding reflections from the walls. The simulations were carried out using a high performance computing, HPC, cluster on 16 CPUs. The time step is 0.0005 s. Details of the CPU run time and computer architecture used by each of the codes for this case are summarised in Table3.

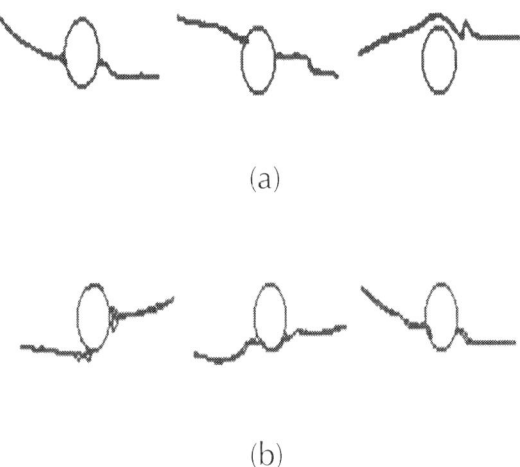

(a)

(b)

62　Dynamics of Offshore Structures

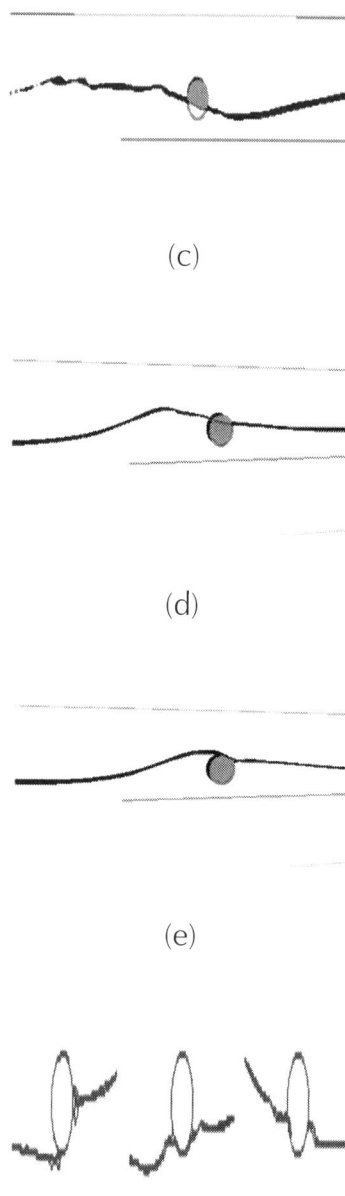

(c)

(d)

(e)

(f)

Investigation of Wave-Structure Interaction Using State... 63

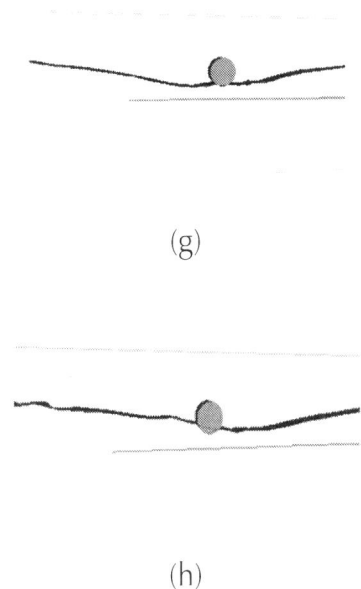

(g)

(h)

Figure 7: Snapshots of surface elevation (a-f: FV and second group: g-l: AMAZON), case. d' = 0.0, A' = 0.5; (a,g) t/T = 0.0, (b,h) t/T = 0.12, (c,i) t/T = 0.36, (d,j) t/T = 0.6, (e,k) t/T = 0.73, (f,l) t/T = 1.0.

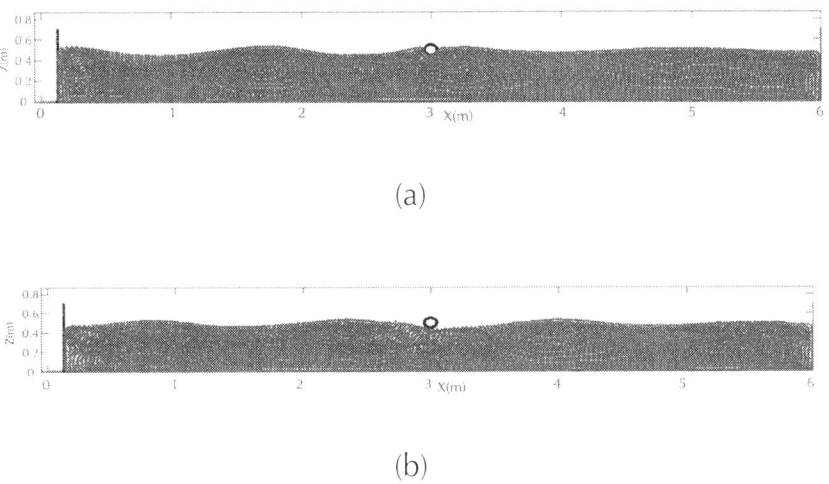

(a)

(b)

Figure 8: Particle configuration for half submerged cylinder (SPH). (a) t/T = 3.5; (b) t/T = 5.0.

For the AMAZON simulations a 2 m × 1.6 m axisymmetric domain is used. The still water level is set to 1.0 m and the initial draught of the cone is 0.148 m. The calculations are performed on a hexahedral grid using an axisymmetric (2D) version of the code with cell sizes of 0.01 × 0.01 m. The time step is 0.00005 s.

The exported vertical forces F_z from the CFD codes are non-dimensionalised using the expression

$$F'(t) = \frac{F_z(t)}{\rho g \pi r^2 A} \quad (8)$$

with ρ being the density of fresh water, g the acceleration due to gravity, r the cone radius at still water level and A the maximum excursion. Also the time is divided by the corresponding period of the central frequency ω_0. The measured relative motion of the water surface is divided though the maximum excursion A = 0.05 m.

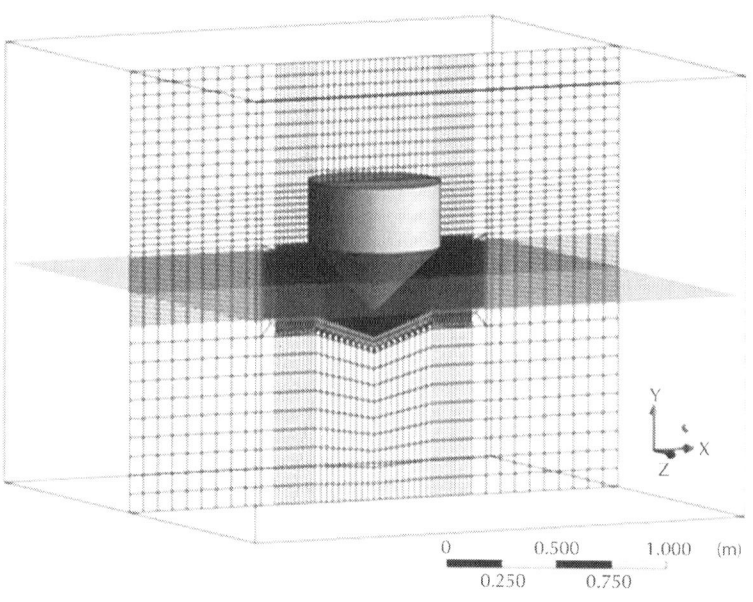

Figure 9: Domain for the oscillating cone case with a section of the mesh at the centreline. The dark region close to the body shows the highly refined mesh, which was necessary to capture the surface jet effects.

Table 3: Details of the NWT used by each code for the oscillating cone case

	CV_FE	AMAZON SC	SPH
Domain dimensions (m)	2.5 * 2.5 * 2.0	2.0 * 1.6 (2D)	4.0 * 4.0 * 1.0
No. of cells/particles	820,000 (3D)	32,000 (2D)	272,000 (3D)
Time step (s)	0.00005	0.00005	0.0001
Convergence error/criteria	0.0001	0.0001	N/A
Computer architecture used	HPCx cluster on 16 1.5 GHz processors	1 processor, 2.5 GHz Macbook Pro computer	1 processor of 1.2 GHz Linux workstation
CPU time: simulation time	3.5 days: 6 s	11 hours: 6 s	1.5 days to 6 s

Figure 10 compares the force data obtained by the three CFD methods with those of the physical experiments in which m = 9. Here, t/T = 0 corresponds to the instant when the vertical impact force is maximum. It is so chosen because different simulations have a different starting time, e.g., for the AMAZON code simulations, t is time from zero and t_o = total time/2 in equation (5), where the total time is defined to be 8T = 5.333 s for the case in which m = 9, and therefore, (t-t_o) is negative. Generally the agreement between numerical prediction and physical experiment is satisfactory. The Eulerian techniques generate little differences especially in the force minima and maxima. AMAZON slightly overestimates the forces and the Navier-Stokes solver underestimates them. The Lagrangian SPH method also agrees well although the troughs are less well resolved.

For the solution of the relative motion of the free surface at the cone, shown in Figure 11, greater differences between the numerical predictions and experiments are evident. Here, the mesh-based AMAZON and CV-FE methods produce results that under-predict the peak values and predict shallower troughs than were measured in the experiment. The particle-based SPH results, however, generally agree very well with measured free surface position at the cone edge, although the predicted troughs are deeper. The difficulty with capturing the free surface close to the cone is due to the occurrence of a jet-effect. This is not apparent for cone cases with low central frequency, where the relative motion of the surface elevation is resolved better.

The grid convergence index (GCI) has been examined for the AMAZON code and results shown in Figure 12. For three levels of mesh resolution (mesh 1 represents the finest mesh with $dx = dy = 0.01$ m, mesh 2 represents $dx = dy = 0.02$ m and mesh 3 represents $dx = dy = 0.04$ m), the RMS of the non-dimensionalized vertical forces are 0.3387, 0.3400 and 0.3475 respectively, from which it can be calculated that the value of GCI_{32} is approximately 0.9% and GCI_{21} is 0.16%. For the SPH code, results for the grid convergence study are shown in Figure 13. Two levels of resolution are tested; particle spacing, $\Delta = 0.02$ m and $\Delta = 0.04$ m. The RMS of the non-dimensionalized vertical forces are 0.281 and 0.258 respectively, from which it can be calculated that the value of GCI_{21} is approximately 0.5%.

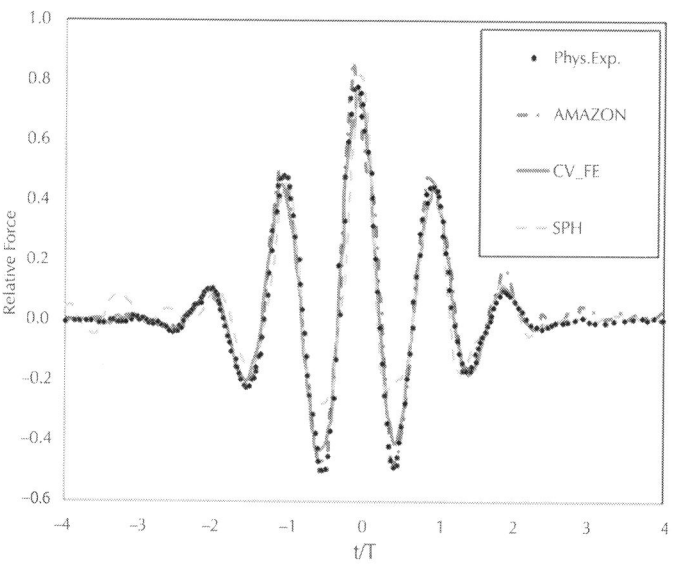

Figure 10: Relative vertical force on cone, m = 9.

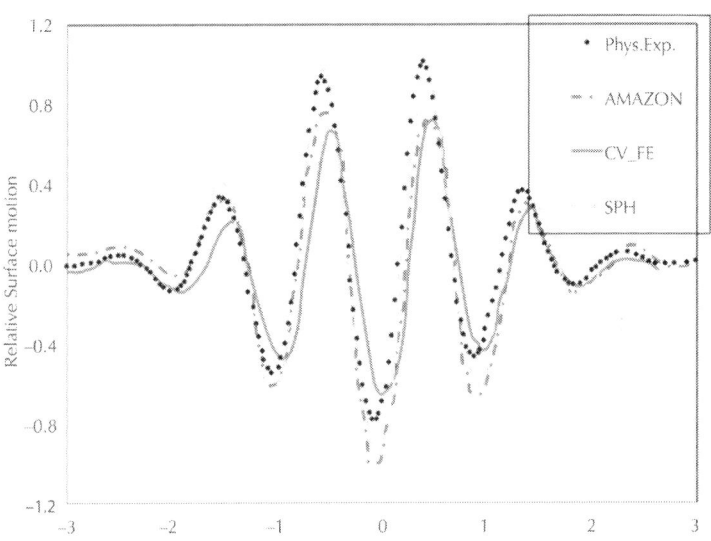

Figure 11: Relative surface motion at cone, m = 9.

Using the CV-FE approach, simulations have been carried out in pairs in order to consider the positive direction cone displacement for a maximum excursion of A = +50 mm and the opposite negative displacement for A = −50 mm. To analyse the nonlinearity in the case, the time histories of the relative surface elevations for the paired tests, m = 9, have been subtracted and summed respectively and divided by 2. This enables results to be broken down into linear and higher order components and compared separately. The half-sum is given by

$$\frac{C+T}{2} \quad (9)$$

and the half-difference by

$$\frac{C-T}{2}, \quad (10)$$

where C and T are the solutions for crest focussed and trough focussed cases. When the half-sum is calculated, the odd frequency components cancel out and the even frequency components remain, and the results of this manipulation on the predicted surface elevations for m = 9 are plotted in Figure 14. Thus the sum line in Figure 14 is composed of second and higher order even frequency components and is dominated by the second order terms; similarly when the half-difference is calculated, the even terms are removed and the odd frequency components remain and are dominated by the linear component, shown in the difference line in Figure 14. For the central circular frequency corresponding to m = 9 the relative water surface elevation clearly contains a higher order component represented by the solid line. Unlike the linear part the higher order component is not symmetric about the mean water line. It oscillates with double the frequency of the linear part around a slightly raised water level.

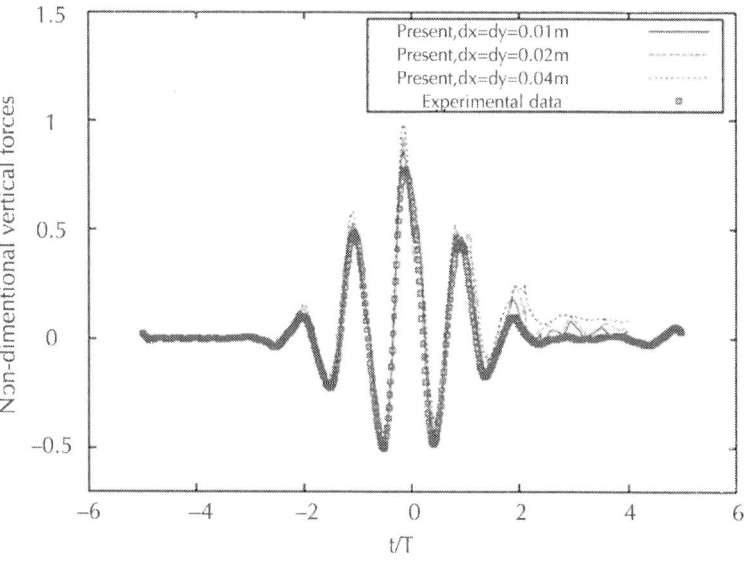

Figure 12: Grid convergence of AMAZON code for relative vertical force on cone.

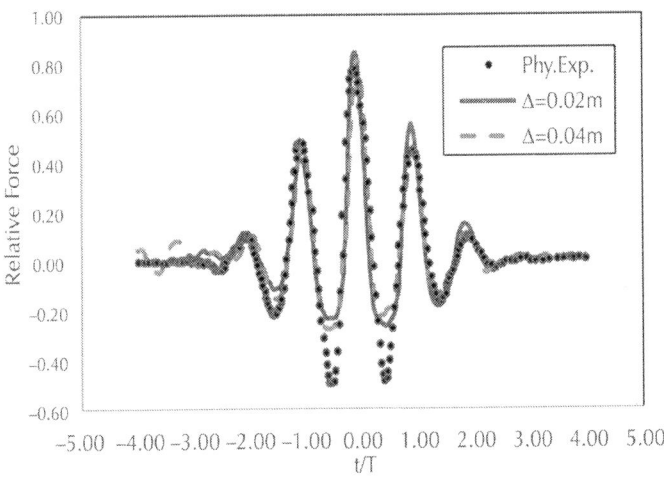

Figure 13: Grid convergence of SPH code for relative vertical force on cone.

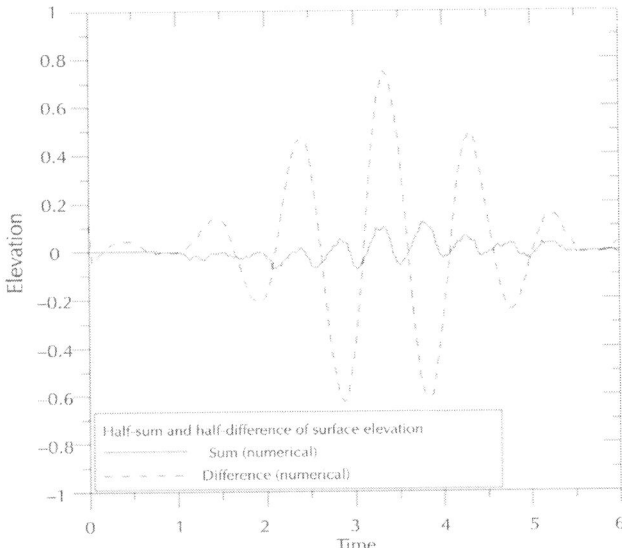

Figure 14: Surface elevation: half-sum and half-difference comparison (m = 9).

Applying the same analysis technique for the vertical forces, results in the plots shown in Figure 15. Here, the higher order parts, i.e. the sum terms, have a double frequency component superimposed on an asymmetric positive component. The total force may be further decomposed into its hydrostatic and hydrodynamic parts. The hydrostatic part results from the buoyancy force, which is subtracted from the total force to obtain the hydrodynamic contribution. Figure 16 shows the non-dimensionalised vertical forces for m = 9 for the maximum excursion negative case, decomposed into dynamic and hydrostatic components. For the lower frequency case, m = 6, the same plot is shown in Figure 17. Here, the hydrodynamic force is a much smaller component of the total force. This is because, due to the higher central circular frequency for larger m, the Keulegan-Carpenter number, K_C, reduces; K_C is given by

$$K_C = \frac{AT_{\omega 0}}{d},$$

(11)

where A is the maximum excursion, $T_{\omega 0}$ the period corresponding to the central frequency and d is the diameter of the cone at the still water level. K_C describes the relationship between the drag forces over the inertia. For lower K_C the inertia dominates the force contribution. This can be seen in the results. For case m = 9, with K_C = 0.11, the dynamic force component, which is related to the inertia of the cone, is more developed than for case m = 6, with K_C = 0.16.

Floating Body Motion: Manchester Bobber in Extreme Waves

Physical tank testing of the Manchester Bobber device at 70th scale was performed in the wave tank at the University of Manchester. It is 18.5 m long, 5 m wide and tests were carried out in a water depth of 0.5 m. The waves are generated using 8 piston type paddles operated using the Edinburgh Designs "OCEAN" interface.

To minimise reflections from the far end wall, a curved surface piercing beach is installed [41] [42].

Here, tests for a single tethered float are reproduced using two Eulerian approaches; the Euler/CartesianCut-Cell and Navier-Stokes/FV Method. A schematic arrangement of the system can be seen in Figure 18, where m_f and m_c are the masses of the float and the counterweight respectively. The horizontal displacement of the float is restricted due to the vertical cables. These are attached to the superstructure and held taut by weights at their ends. In the physical experiments vertical displacements are deduced from the angular displacement of the pulley ω_p. During all tests no power was taken off the system and the friction in the pulley support is negligible. The cables are assumed to be stiff and inextensible. Further details on the experimental system are given by Stallard et al. [41] and Weller et al. [42] for a different float form and mass.

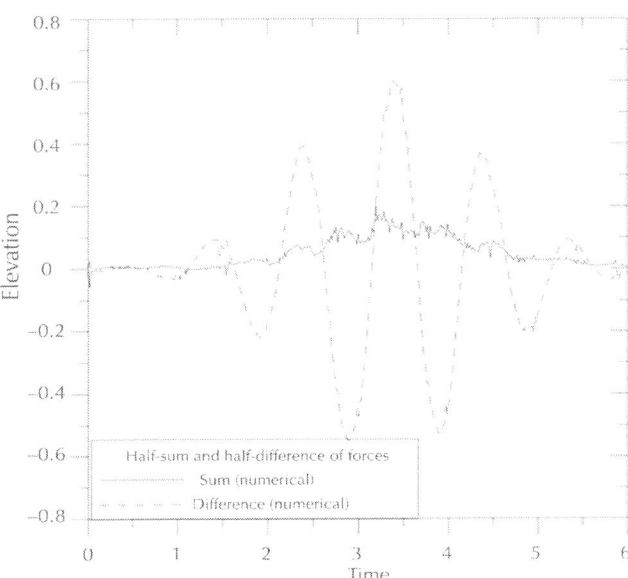

Figure 15: Non-dimensionalised vertical forces: Half-sum and halfdifference comparison (m = 9).

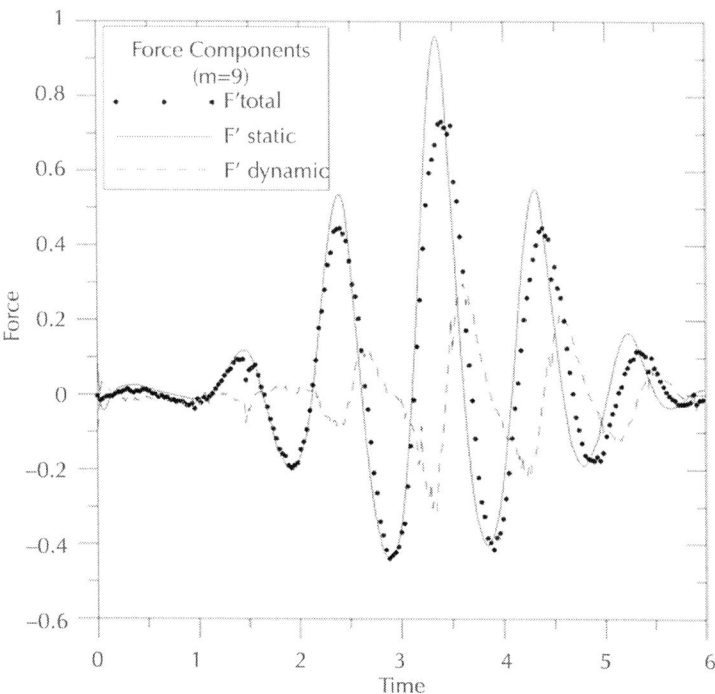

Figure 16: Non-dimensionalised force components for m = 9: maxmum excursion negative.

For the simulation of the mechanical system in CFD it is necessary to know the relationship between the two accelerated bodies, i.e. the float and the counter weight. The reason for this is that the CFD codes do not model the pulley system and the counter weight directly, and so these are approximated using additional body forces. The free body diagram as seen in Figure 19 is used to find the unknown tension forces in the cable T_1 and T_2 and the acceleration of the system.

Investigation of Wave-Structure Interaction Using State... 73

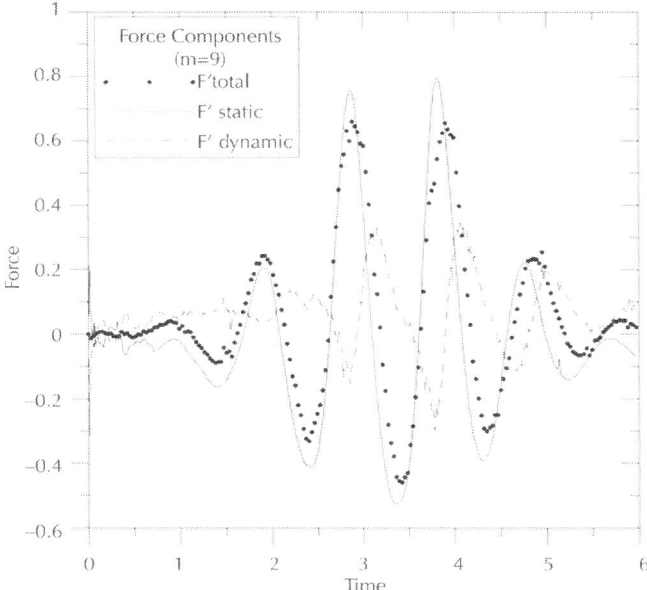

Figure 17: Non-dimensionalised force components for m = 6: maximum excursion positive.

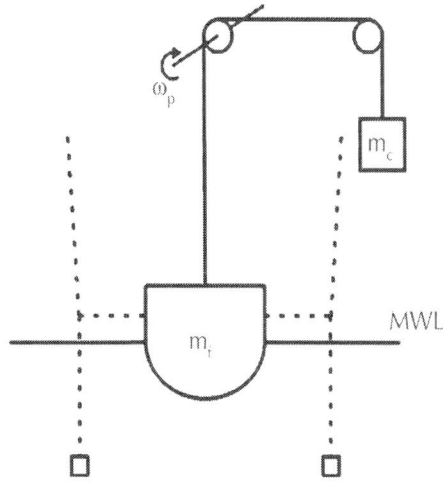

Figure 18: Geometry.

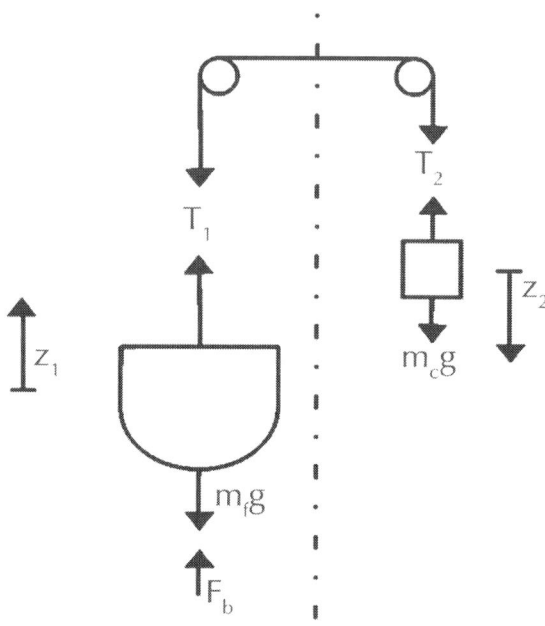

Figure 19: Force analysis.

For the left hand side of system, representing the float, the force equilibrium is achieved when

$$m_f \ddot{z}_1 = -m_f g + T_1 + F_b \quad (12)$$

with m_f being the mass of the float, the positive upward acceleration of the system, g is gravity, T_1 represents the tension force in the cable and F_b is the buoyancy force. For the counterweight on the left hand side, the positive z-direction is downward, denoted z_2, and the force equilibrium can be written as

$$m_c \ddot{z}_2 = m_c g - T_2 \quad (13)$$

where m_c is the mass of the counterweight and T_2 the tension force in the cable. When moving the float vertically by a distance of z_1 the counterweight covers the same distance, from which follows that $Z_1 = Z_2 \Rightarrow \dot{Z}_1 = \dot{Z}_2 \Rightarrow \ddot{Z}_1 = \ddot{Z}_2$. Furthermore, the relationship between the tension forces can be written as.

The two unknowns of the system; the acceleration of the float and the tension force T, can then be written as

$$\ddot{z} = \frac{(m_c - m_f)g + F_b}{m_f + m_c} \tag{14}$$

and

$$T = -m_c \frac{(m_c - m_f)g + F_b}{m_f + m_c} + m_c g \tag{15}$$

In the computational approach F_b is calculated from the integrated pressures on the float surface and thereby known at any time. The simulations are run using AMAZON and the FV solver with two alternative approaches to representing the effect of the float, pulley and counterweight arrangement. These are identified as case A and B and are defined in Table 4, in which the mass of the float and the applied tension force relating to **Figure 19** are given.

For the numerical simulations NewWave focusing is used to generate the extreme wave [43] The concept of wave focusing is to generate several waves of relatively small amplitudes and different periods. These waves interact and constructive interfere to build up a localised extreme wave, larger than any individual wave created at the paddle, focussed at a specified position and time in the tank. For each wave component n the amplitude a_n is defined as

$$a_n = A \frac{S_n(f) \Delta f}{\sum_N S_n(f) \Delta f} \tag{16}$$

where $S_n(f)$ is the spectral density, Δf is the frequency step depending on the number of wave components and bandwidth and A is the target linear amplitude of the focussed wave. Thus, the amplitude of every spectral component in the NewWave group scales as the power density within that frequency band in the assumed sea-state. Equivalently, NewWave is simply the scaled auto-correlation function corresponding to a specified frequency spectrum such as the one obtained on the measured surface elevation time history at the location of the float without the float being in place during the physical tank tests.

The waves are generated using a velocity inlet at the left hand boundary of the numerical wave tank (NWT) at x = 0. Here, the surface elevation of the wave group is prescribed using the same techniques as described for the regular waves in Section 3.1 by specifying the vertical location of the water volume fraction of 0.5. The horizontal and vertical velocity components derived from NewWave theory are applied at the inlet boundary, such that t_0 and x_0 are the chosen focus time and location in the tank, here set to 4.6 s and 3.5 m respectively. N is the number of wave components, here 15. The numerically reproduced wave without the float in place using AMAZON and the FV solver can be seen in Figure 20 together with the measured elevation. Both numerical codes reproduce the focussed wave well. The FV solver predicts the asymmetry of the peak and trough values either side of the main peak particularly well, and the double peak evident in the experiment at the preceding peak is also predicted with this method. However, the depth of the preceding trough is slightly over predicted and there are some unphysical oscillations in the time history predicted for the first three seconds of the simulation. The AMAZON code predicts the peak of the main wave well, but there is a time lag in the prediction of the following wave and some differences in the position of the preceding wave. Details of the CPU run time and computer architecture used by each of the codes for this case are summarized in Table5

Figures 21 (Case A) and 22 (Case B) show the comparison of predicted and measured vertical translation of the float as it interacts with the extreme wave. The set-up, including details of the mesh and the applied forces as well as an example of the NWT showing the free surface at a point in time as the focused wave approaches the float are shown in Figure 23 and in Figure 24photographs of the experiment in the wave tank at Manchester

University are given to show the large horizontal motions observed in addition to vertical motion of the float. It should be noted that these numerical simulations are restricted to vertical motion only and are therefore restrained in horizontal motion and rotation. This is a current limitation of the software and would explain the differences between the predicted and measured vertical motion of the float.

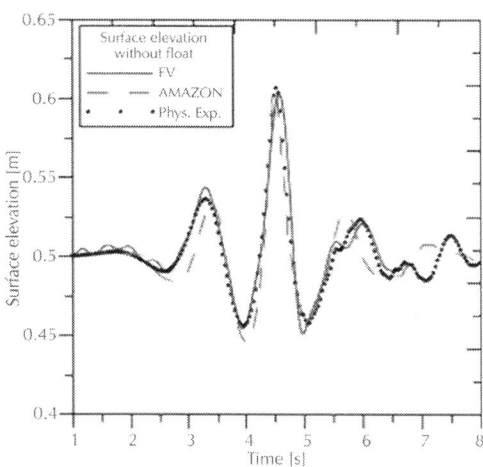

Figure 20: Surface elevations at location of float without the float being in place.

Case A, in which a constant tension force of $T = m_c g$ and the mass of float, $m_f - m_c$ are applied, predicts significant damping of the vertical motion response after the focused wave has passed although the initial motion of the float is predicted well. In case B, the tension force varies with the acceleration of the float calculated within the solution as, with the mass of float, m_f. This leads to a much better prediction of the vertical motion in general, and whereas the response is slightly over-predicted as the extreme wave interacts with the float, its response is much better predicted following the extreme event. Thus the solution is much improved when the tension force includes the instantaneous acceleration of the float rather that assuming it is constant as in case A.

In case C, another Bobber shape is tested in the wave tank at Manchester University. In this test, the geometry of the Bobber is a flat-bottomed cylinder of radius 0.074 m with a corner radius of 0.033 m. The vertical sides extend to 0.085 m above the flat base and a conical upper surface with a 30 degree base angle decreases the radius of the geometry to 0.025 m at the upper cylindrical section (see Figure 25). Figure 26 shows the surface mesh of the flat-base Bobber. The outer dimensions of the numerical domain are and the water depth is 0.35 m. Two different particle densities are used and results presented in Figures 26 and 27. The number of particles in the coarse mesh is 118,000 with particle spacing of 0.04 m and the fine mesh is 918,000 with particle spacing of 0.02 m.

According to the experimental tests by Stallard et al. [41] and Weller et al. [41], the float mass m_f = 2.1 kg and the counterweight mass m_c = 1.0 kg in case C. Figure 27 and Figure 28 show the comparison of SPH results and experimental data for the device response, using uniform particle mass and solving the dynamics in one and six degrees of freedom using two particle resolutions (Δx = 0.04 m and Δx = 0.02 m). The maximum wave amplitude produces the second peak in the device-response profile at t = 4.6 s or t/T_p = 3.2. The results are in agreement in terms of phase and magnitude. However, the SPH result for the coarse simulations seems to be oscillatory because the number of fluid particles interacting with the device is small, whereas for the finer resolution the response is considerably smoother and all the peaks are reproduced by the SPH results. In order to achieve a smoothed profile of device response, the number of fluid particles around the device needs to be considerably greater. Clearly, the SPH results for the system with six degrees of freedom are smoother and in better agreement (Figure 28) than the system with one degree of freedom.

Table 4: Properties of manchester bobber cases

	A	B	C
Mass of float	$m_f - m_c$	m_f	$m_f - m_c$

Vertical Motion	yes	yes	yes
Horizontal Motion	-	-	yes
Tension force T	$m_c g$	$m_c(g - \ddot{Z})$	$m_c(g - \ddot{Z})$
Solvers	FV/AMAZON	FV	SPH

Table 5: Details of the NWT used by each code for the manchester bobber case

	FV	AMAZON SC	SPH
No. of cells/particles	530,000	287,550	118,000
Time step (s)	0.0005	0.00005	0.0001
Convergence error/criteria	0.0001	0.0001	N/A
Computer architecture used	HPCx cluster on 8 processors, 2.5 GHz, 2 GB per node with Infiniband	1 processor of 600 MHz NEC vector computer	16 processors of 1.2 GHz Linux workstation
CPU time: simulation time	3 days: 8 s	20 days: 8 s	1 day: 8 s

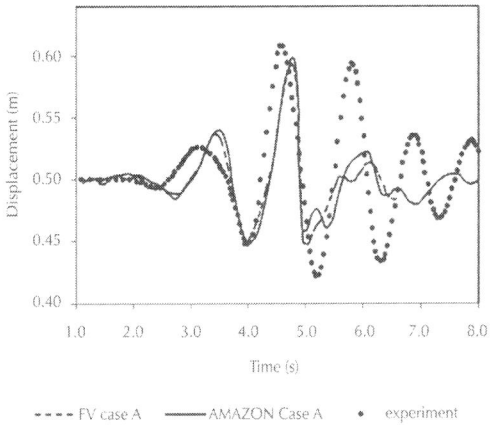

Figure 21: Translation of float (case A).

80   Dynamics of Offshore Structures

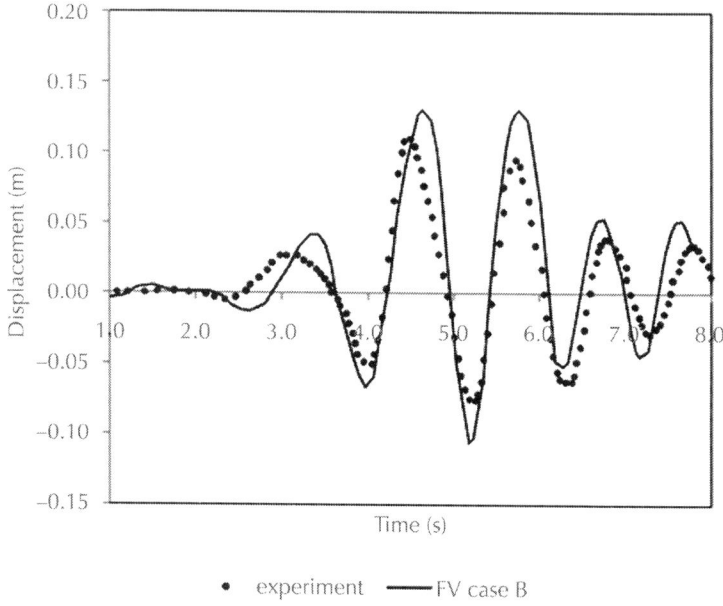

Figure 22: Translation of float (case B).

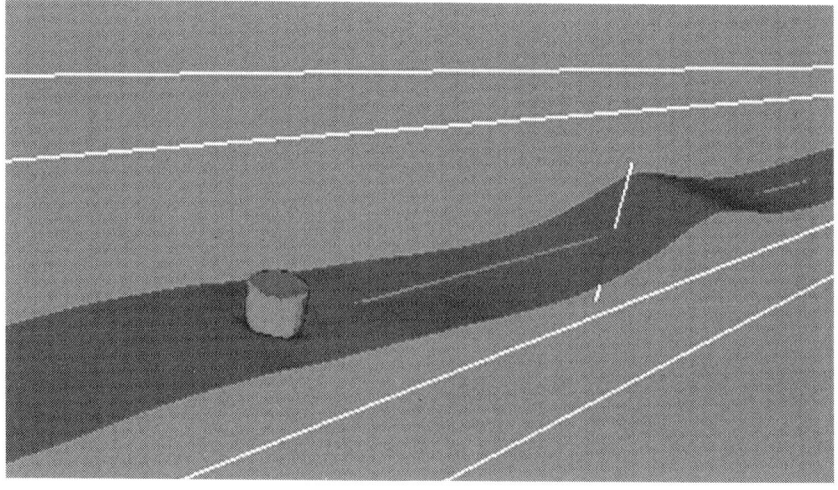

(a)

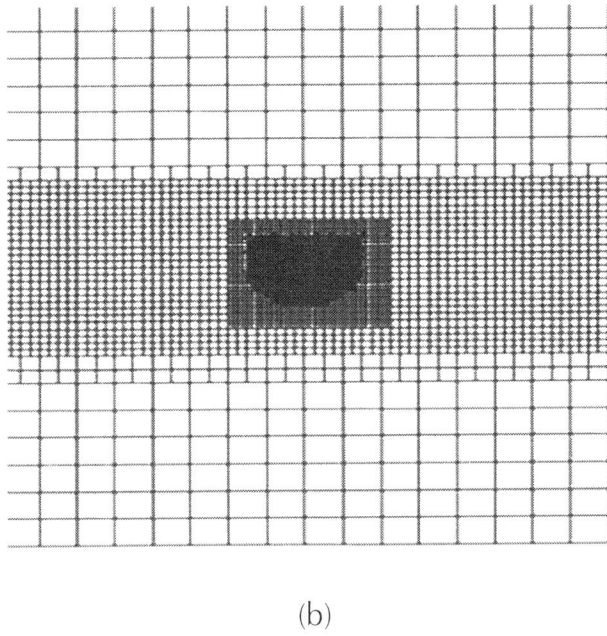

(b)

Figure 23: Details of the mesh and free surface at a point in time as the focused wave approaches the float.

(a)

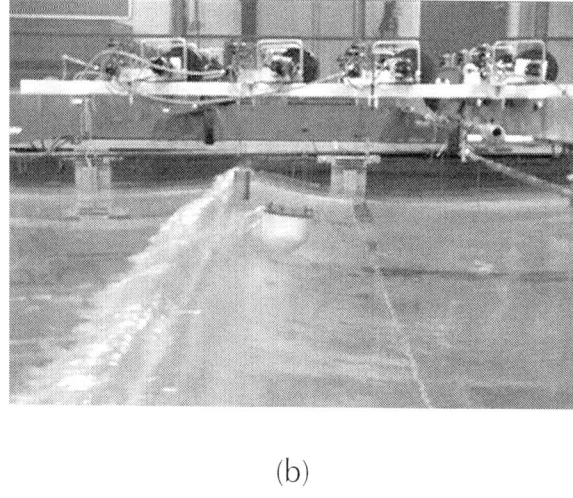

(b)

Figure 24: Position of the float in the physical experiments for two instances in time. Left: approaching large wave from the right. Right: the breaking wave has just passed the float, which is airborne. Counterweight and float have moved significantly from the position shown on the left.

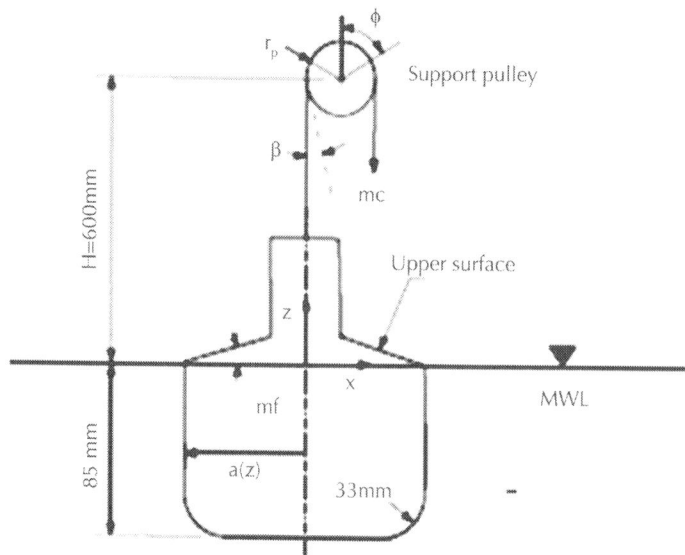

Figure 25: Case C geometry.

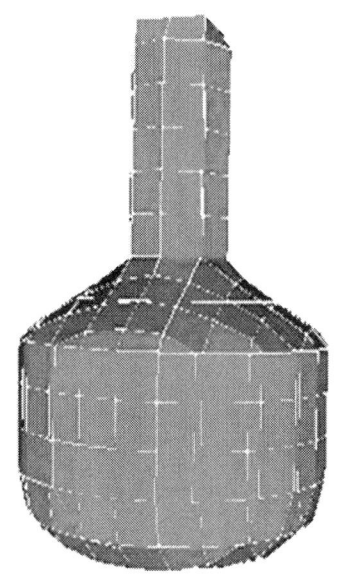

Figure 26: Case C bobber with flat-bottom cylinder.

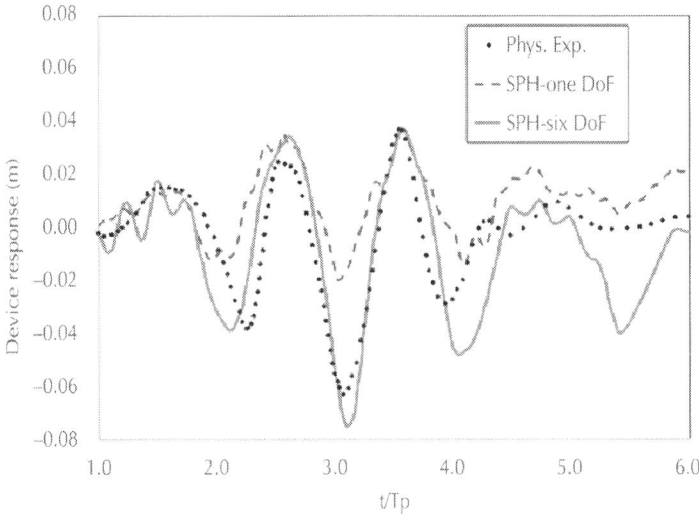

Figure 27: Comparison of SPH result and experimental data for the device response using uniform particle mass, (x = 0.04 m), case C.

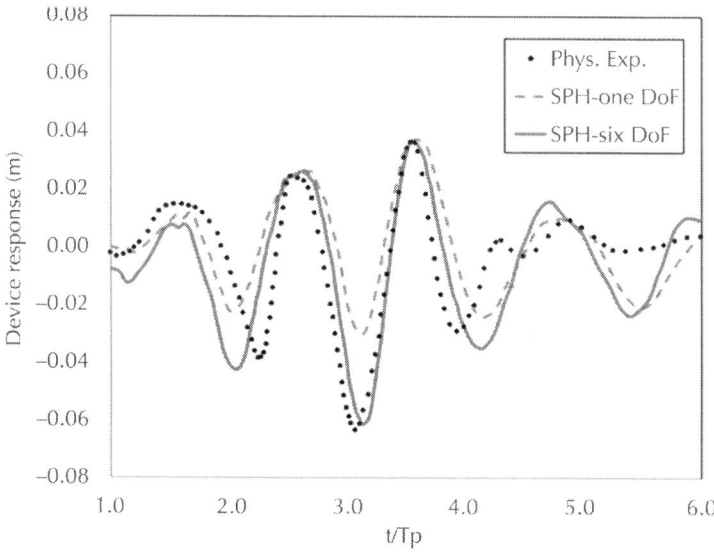

Figure 28: Comparison of SPH result and experimental data for the device response using uniform particle mass, (x = 0.02 m), case C.

DISCUSSION AND CONCLUSIONS

Three non-linear wave-structure interaction cases have been modelled using state-of-the-art CFD techniques that include three mesh-based techniques solving the Euler and Navier-Stokes equations, and one particle method. The cases increase in complexity, starting with a fixed horizontal cylinder in regular waves. The second case involves rigid vertical motion of a cone, which is forced to oscillate and radiate waves at the surface in still water. The last experiment shows the ability of the CFD methods to cope with highly non-linear wave and floating body interaction. Furthermore, in the last test the body not only interacted with waves, which was modelled sufficiently for smaller waves by Hadzic et al. [4] , but also with an attached mass, which is represented in the numerical simulations here by a body force.

The use of CFD for wave-structure interaction is usually only considered for situations in which assumptions such as inviscid and irrotational flow, linear wave theory and small body motion do not apply. In these situations, the detailed information provided through CFD modelling is of interest and must be considered alongside the accuracy of the results given by CFD, the computational costs and the flexibility in setup and problem definition. Generally the accuracy of the results increases with the number of cells/particles in the domain, and it is possible to capture highly non-linear effects like spray at the cost of larger run-times. Here, fluid forces, surface elevations and motions are predicted and compared with experimental data. For a fixed structure, the forces are well resolved by all methods applied and non-linear surface interaction was also captured successfully.

Body motion has been modelled in two ways. First, a cone-shaped body was forced at the free surface to radiate waves. The difficulties of this case for the mesh methods lay in the presentation of the body and its small maximum displacement. For free surface flow, the mesh should normally use hexahedral cells, as stated in Westphalen et al. [39] . However, for modelling the axisymmetric cone this is not beneficial, as either the cells are distorted (when using a body fitted grid) or arbitrarily shaped, which both introduce numerical errors. These increase when the mesh moves according to rigid body motion. In SPH this problem is non-existent. All results, however, showed good agreement.

The second case involving rigid body motion is the heave response of a floating body to interaction with an extreme focussed wave. Here, the results for the CFD packages differ. While the Navier-Stokes/FV solver computed the translation of the body with good agreement when the tension force was modelled by the gravitational component only, it overestimates the motion of the float when the tension force also includes the time-varying body acceleration. The problem of overestimation of the motion due to using the unsteady part of the tension force, needs to be resolved, and may well be due to the numerical solutions being limited to heave motion, whereas the experiments clearly show large surge motions as seen in Figure

24. The SPH certainly show better agreement with experiment with 6 degrees of freedom than with heave alone.

The extreme wave loading presented in this paper makes use of a NewWave design wave to reduce the run times. In reality, irregular waves excite the body continuously and these long-term irregular waves are needed to model the final device with mooring and PTO. The computational cost for such runs would increase significantly, not only for running the simulations but also for storing the results. The parallel capability of all the CFD packages used for this work makes it possible to reduce the wall clock time significantly, and in some cases symmetry can be employed to achieve further reductions. As the simulations presented here were performed on different platforms, direct comparisons are difficult. To give an example for the Navier-Stokes FV solver: The final simulation of 8 s of the floating body needed 5 days on 8 partitions using a desktop workstation (16 GB RAM, 2.5 GHz). The same simulation on a cluster took 3 days (2.5 GHz, 2 GB per node, Infiniband).

In this paper the use of CFD for wave loading on WECs is considered using four different CFD techniques. From the results presented it can be concluded that CFD is a versatile tool, capable of solving highly complex and non-linear wave-structure interaction problems. Each of the CFD methods considered predicted the wave structure interaction well and in generally good agreement with experimental results. The jet flow developed in the oscillating cone case is challenging for the Eulerian mesh based approaches as the resolution of the jet is limited by the grid size, and SPH is not limited in the same way and is better able to capture the jet.

ACKNOWLEDGEMENTS

The authors would like to acknowledge the support of the Engineering and Physical Sciences Research Councilfunded under the project title "Extreme Wave Loading on Offshore Wave Energy Devices using CFD: a Hierarchical Team Approach" (Grant No. EP/D077508).

Furthermore, many thanks go to Dr. Kevin Drake of University College London (UCL) for providing the high quality physical test data used for the cone simulations.

REFERENCES

1. WAMIT Inc. (2006) User Manual, Versions 6.4, 6.4 PC, 6.3, 6.3S-PC.
2. Delhommeau G. (1993) Seakeeping Codes AQUADYN and AQUAPLUS. In: Proceedings of the 19th WEGEMT SCHOOL on Numerical Simulation of Hydrodynamics: Ships and Offshore Structures, Nantes.
3. Cummins, W.E. (1962) The Impulse Response Function and Ship Motions. Schiffstechnik, 47, 101-109.
4. Hadzic, I., Hennig, J., Peric, M. and Xing-Kaeding, Y. (2005) Computation of Flow-Induced Motion of Floating Bodies. Applied Mathematical Modelling, 29, 1196-1210. http://dx.doi.org/10.1016/j.apm.2005.02.014
5. Dixon, A.G., Greated, C.A. and Salter, S.H. (1979) Wave Forces on Partially Submerged Cylinders. Journal of the Waterway, Port, Coastal and Ocean Devision, 105, 421-438.
6. Drake, K.R., Eatock Taylor, R., Taylor, P.H. and Bai, W. (2008) On the Hydrodynamics of Bobbing Cones. Ocean Engineering 36, 1270-1277. http://dx.doi.org/10.1016/j.oceaneng.2009.07.007
7. CD-Adapco (2009) STAR CCM+Version 4.04.011. London.
8. Tu, J. Yeoh, G. and Liu, C. (2008) Computational Fluid Dynamics: A Practical Approach. 2nd Edition, Butterworth-Heinemann, Oxford.
9. Patankar, S.V. (1980) Numerical Heat Transfer and Fluid Flow. Taylor & Francis, London.
10. Patankar, S.V. and Spalding, D.B. (1972) A Calculation Procedure for Heat, Mass and Momentum Tranfer in Three- Dimensional Parabolic Flows. International Journal

of Heat and Mass Transfer, 15, 1787-1806. http://dx.doi.org/10.1016/0017-9310(72)90054-3
11. Hirt, C.W. and Nichols, B.D. (1981) Volume of Fluid (VOF) Method for the Dynamics of Free Boundaries. Journal of Computational Physics 39, 201-225. http://dx.doi.org/10.1016/0021-9991(81)90145-5
12. Ubbink, O. (1997) Numerical Prediction of Two Fluid Systems with Sharp Interfaces. PhD: 138 Department of Mechanical Engineering, Imperial College of Science, Technology & Medicine, London.
13. Greaves, D.M. (2004) A Quadtree Adaptive Method for Simulating Fluid Flows with Moving Interfaces. Journal of Computational Physics, 194, 35-56. http://dx.doi.org/10.1016/j.jcp.2003.08.018
14. Ferziger, J.H. and Peric, M. (2001) Computational Methods for Fluid Dynamics. Springer, Heidelberg.
15. Ansys, I. (2006) ANSYS CFX-Solver Theory Guide. Canonsburg. http://www.ansys.com/staticassets/ANSYS/staticassets/resourcelibrary/brochure/ansys-cfx-tech-specs.pdf
16. Baliga, B.R. and Patankar, S.V. (1980) A New Finite-Element Formulation for Convection-Diffusion Problems. Numerical Heat Transfer, Part A, 3, 393-409.
17. Baliga, B.R. and Patankar, S.V. (1983) A Control Volume Finite-Element Method for Two-Dimensional Fluid Flow and Heat Transfer. Numerical Heat Transfer, Part A, 6, 245-261.
18. Barth, J.T. and Jesperson, D.C. (1989) The Design and Application of Upwind Schemes on Unstructured Meshes. American Institute of Aeronautics and Astronautics (AIAA), Reno.
19. Zwart, P.J., Scheuerer, M. and Bogner, M. (2003) Free Surface Modelling of an Impinging Jet. ASTAR International Workshop on Advanced Numerical Methods for Multidimensional Simulation of Two-Phase Flow, Garching, 15-16 September 2003.

20. Zwart, P.J. (2005) Numerical Modelling of Free Surface and Cavitating Flows. VKI Lucture Series, Ansys Canada Ltd Canada, 25.
21. Rhie, C.M. and Chow, W.L. (1982) A Numerical Study of the Turbulent Flow Past an Isolated Airfoil with Trailing Edge Separation. AIAA/ASME 3rd Joint Thermophysics, Fluids, Plasma and Heat Transfer Conference, St. Louis, 7-11 June 1982.http://dx.doi.org/10.2514/6.1982-998
22. Monaghan, J.J. (2005) Smoothed Particle Hydrodynamics. Reports on Progress in Physics 68, 1703-1759. http://dx.doi.org/10.1088/0034-4885/68/8/R01
23. Vila, J.P. (1999) On Particle Weighted Methods and Smoothed Particle Hydrodynamics. Mathematical Models and Methods in Applied Sciences 9, 161-209.
24. Guilcher, P.M., Ducorzet, G., Alessandrini, B. and Ferrant, P. (2007) Water Wave Propagation Using SPH Models. Proceedings of 2nd International SPHERIC Workshop, Madrid, 23-25 May 2007, 119-124.
25. Toro, F. (2001) Shock-Capturing Methods for Free-Surface Shallow Flows. John Wiley & Sons LTD, Hoboken.
26. Hirsch, C. (1990) Numerical Computation of Internal and External Flows, volume 2: Computational Methods for Inviscid and Viscous Flows, Wiley, Hoboken.
27. Chatkravathy, S.R. and Osher, S. (1983) High Resolution Applications of the Osher Upwind Scheme for the Euler Equations, AIAA Paper 83-1943. Proceedings of AIAA 6th Comutational Fluid Dynamics Conference, Danvers, July 1983, 363-373.
28. SPHysics. (2010) HYPERLINK. http://wiki.manchester.ac.uk/sphysics.
29. Monaghan, J.J. and Kos, A. (1999) Solitary Waves on a Creatan Beach. Journal of Waterway, Port, Coastal and Ocean Engineering, 125, 145-154.http://dx.doi.org/10.1061/(ASCE)0733-950X(1999)125:3(145)

30. Chorin, A.J. (1968) Numerical Solution of the Navier-Stokes Equations. Mathematics of Computation, 22, 745-762.
31. Cummins, S.J. and Rudman, M. (1999) An SPH Projection Method. Journal of Computational Physics, 152, 584-607.
32. Lind, S., Xu, R., Stansby, P. and Rogers, B. (2012) Incompressible Smoothed Particle Hydrodynamics for Free-Surface Flows: A Generalised Diffusion-Based Algorithm for Stability and Validations for Impulsive Flows and Propagating Waves. Journal of Computational Physics, 231, 1499-1523. http://dx.doi.org/10.1016/j.jcp.2011.10.027
33. Skillen, A., Lind, S., Stansby, P. and Rogers, B. (2013) Incompressible Smoothed Particle Hydrodynamics (SPH) with Reduced Temporal Noise and Generalised Fickian Smoothing Applied to Body-Water Slam and Efficient Wave-Body Interaction. Computer Methods in Applied Mechanics and Engineering, 265, 163-173.http://dx.doi.org/10.1016/j.cma.2013.05.017
34. Xu, R., Stansby, P.K. and Laurence, D. (2009) Accuracy and Stability in Incompressible SPH (ISPH) Based on the Projection Method and a New Approach. Journal of Computational Physics, 228, 6703-6725. http://dx.doi.org/10.1016/j.jcp.2009.05.032
35. Qian, L., Causon, D.M., Mingham, C.G. and Ingram, D.M. (2006) A Free Surface Capturing Method for Two Fluid Flows with Moving Bodies. Proceedings of the Royal Society A: Mathematical, Physical and Engineering Science 462, 21-42.
36. Causon, D.M., Ingram, D.M., Mingham, C.G., Yang, G. and Pearson, R.V. (2000) Calculation of Shallow Water Flows Using a Cartesian Cut Cell Approach. Advances in Water Resources 23, 545-562. http://dx.doi.org/10.1016/S0309-1708(99)00036-6
37. Causon, D.M., Ingram, D.M. and Mingham, C.G. (2001) A Cartesian Cut Cell Method for Shallow Water Flows with Moving Boundaries. Advances in Water Resources 24, 899-911.http://dx.doi.org/10.1016/S0309-1708(01)00010-0

38. Westphalen, J. (2010) Extreme Wave Loading on Offshore Wave Energy Devices Using CFD. PhD Thesis, University of Plymouth, Plymouth 73-75.
39. Westphalen, J., Greaves, D., Williams, C.J.K., Zang, J. and Taylor, P. (2008) Numerical Simulation of Extreme Free Surface Waves. Proceedings of the 18th International Offshore and Polar Engineering Conference, Vancouver, 6-11 July 2008.
40. Rogers, B.D. and Dalrymple, R.A. (2008) SPH Modelling of Tsunami Waves. In: Advances in Coastal and Ocean Engineering Vol. 10, Advanced Numerical Models for Tsunami Waves and Runup. W. Scientific, Singapore.
41. Stallard, T., Weller, S.D. and Stansby, P.K. (2009) Limiting Heave Response of a Wave Energy Device by Draft Adjustment with Upper Surface Immersion. Applied Ocean Research, 31, 282-289. http://dx.doi.org/10.1016/j.apor.2009.08.001
42. Weller, S.D., Stallard, T.J. and Stansby, P.K. (2012) Experimental Measurements of the Complex Motion of a Suspended Axisymmetric Floating Body in Regular and Near-Focused Waves. Applied Ocean Research, 39, 137-145.
43. Taylor, P.H. and Williams, B.A. (2004) Wave Statistics for Intermediate Depth Water-NewWaves and Symmetry. Journal of Offshore Mechanics and Arctic Engineering, 126, 54-59. http://dx.doi.org/10.1115/1.1641796

Chapter 3

Advantages of the Green Solid State FSW over the Conventional GMAW Process

Hasan I. Dawood[1,2], Kahtan S. Mohammed[1], and Mumtaz Y. Rajab[3]

[1]School of Materials Engineering, Universiti Malaysia Perlis, Taman Muhibah-Jejawi, 02600 Arau, Perlis, Malaysia
[2]Chemical Engineering Department, AL-Qadisiyah University, Iraq
[3]Mechanical Engineering Department, AL-Qadisiyah University, Iraq

ABSTRACT

The present work is an experimental comparison between the friction stir welding (FSW) and the conventional gas metal arc welding (GMAW) in joining of Al alloys. Two sets of 3 mm thick aluminum strip pairs were friction stir welded in a regular butting joint configuration. Two rotational speeds of 1750 rpm and 2720 rpm were utilized to perform the FSW process. The axial force and the transverse speed were kept constant at 6.5 KN and 45 mm/min, respectively. Cylindrical tool shoulder and pin geometry were selected. Strip pairs of other similar sets were butt jointed using the conventional GMAW. The welding quality, power input, and macrostructure and microstructure of the butted joints were examined. The types of the fumes and the amount of the released gases were measured and compared. The results showed that the solid state FSW is green, environment-friendly, and of superior welding properties compared to the conventional GMAW.

INTRODUCTION

FSW is a solid state welding process which can afford a high quality of welds even for materials that are unmanageable with conventional welding such as aluminum. It is a clean, environment-friendly, and nonharmful process as it is accompanied by neither an arc formation and radiation nor toxic gas emission. It has low heat input and almost no welds finishing costs [1, 2]. FSW gives several advantages over other welding techniques for joining various alloys, especially light alloys [3]. Owing to certain properties such as light weight, high strength to weight ratio, and good corrosion resistance, aluminum alloys are used in wide applications, including aerospace, automobile industries, shipbuilding, and train and tram wagons. They are also used in offshore structures and bridge construction. So far FSW has been widely applied to welding of low melting alloys which are difficult to be joined by any other conventional fusion welding such as Al-Li 2195 alloy [4].

The process and terminology of FSW are schematically represented in Figure 1.

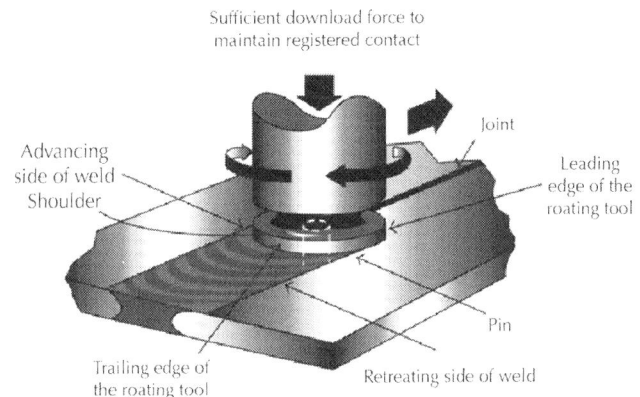

Figure 1: The schematic representation of FSW process.

During FSW, heat input for welding is provisioned by the rubbing action of the tool shoulder with the top surface of the welded piece and by the plastic dissipation of the mechanical energy generated by the tool pin [5]. However, Roy et al. [6] and Hirata et al. [7] reported that the heat flow from the pin is relatively small compared to flow from the shoulder and thus it can be ignored. The heat input creates a softened plasticized metal around the tool and facilitates its transverse movement along the joint line. The plasticized metal is mixed, sheared, and extruded around the rotating tool pin in the vertical direction under the applied axial force. Eventually, the plasticized metal is forged by the contact of the tool shoulder and the pin resulting in a solid phase bond between the two pieces. During the welding process, advancing side is the side which the velocity vector of rotational speed is in the same direction with the welding speed and the other side which represent the retreating side. The FSW process leads to the appearance of thermomechanically affected zone (TMAZ), a heat affected zone (HAZ), and a nugget zone (NZ) which is of a vase-like shape in the central part of the TMAZ. The main FSW parameters that determine the quality of the welded joint are the tool plunge force, the tool rotation speed, and the travel or traverse speed [8].

GMAW is a conventional arc welding process [9]. Schematic representation of this process is shown in Figure2 [9]. Throughout welding, heat input is an important characteristic; it influences the cooling rate and consequently the mechanical properties and the metallurgical structure of both the WZ and the HAZ [10]. GMAW welding process is performed at temperatures above the melting temperature of the workpiece. The higher temperature process needs higher power and can induce lots of defects in the welded piece such as distortion, cracking, and higher residual stresses which result in inferior mechanical properties [8].

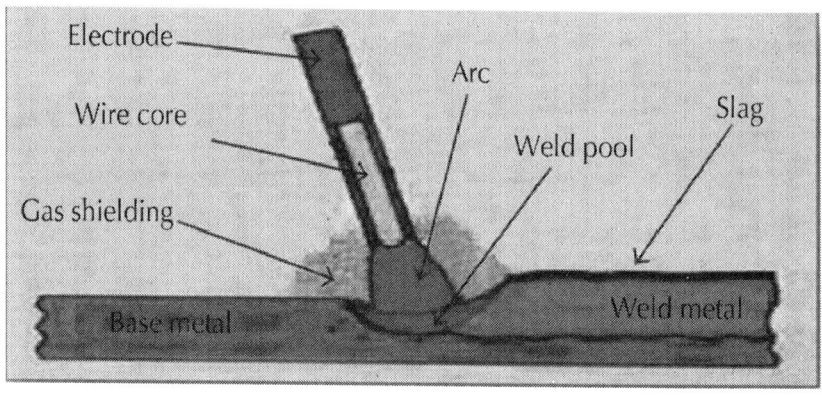

Figure 2: The schematic diagram of the welding process by GMAW.

In this study, two sets of 3 mm thick aluminum strip pairs were friction stir welded together at rotational speeds of 1750 rpm and 2720 rpm. Welding process for another similar set of pairs was repeated using the conventional GMAW. Several papers are found comparing FSW with different conventional fusion arc welding processes; few of them are on welding of Al alloy [1, 11, 12]. The aim of this paper is to study the influence of FSW and GMAW on microstructure evolution and the quality of the Al-Al weld joints and to assess their impact on the environment.

MATERIALS AND EXPERIMENTAL PROCEDURES

3 mm thick 1030 Al strips with a nominal chemical composition of 0.114% Si, 0.405% Fe, 0.084% Cu, 0.011% Mn, 0.032% Mg, 0.017% Ti and Al balance were used in this investigation. The FSW and the GMAW joints were of butt type. They were performed on a milling machine. Prior to welding, the metal pairs were cleaned using acetone to remove any grease and stains that may affect the quality of welding. The metal pair was fixed by clamps to prevent any movement during the welding process as shown in Figure 3.

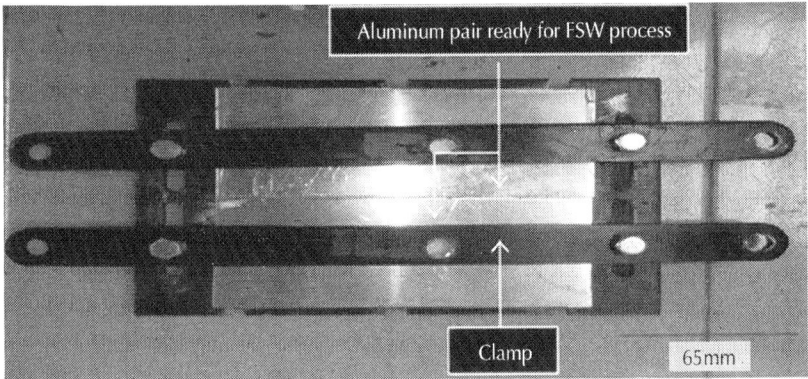

Figure 3: Photograph of the clamped Al strips.

The horizontalness of the clamped pair on the milling machine was assured by horizontal level indicator for all Al-Al weld joints. Figure 4 shows that the utilized tool dimensions (shoulder and pin) were made of medium carbon steel (0.424% C, 0.727% Mn, 0.013% P, and 0.17% Si, and Fe balance). The tool was heat-treated and quenched to RC 56. The vertical milling machine used in this work was about 3 horsepower (2237 watts).

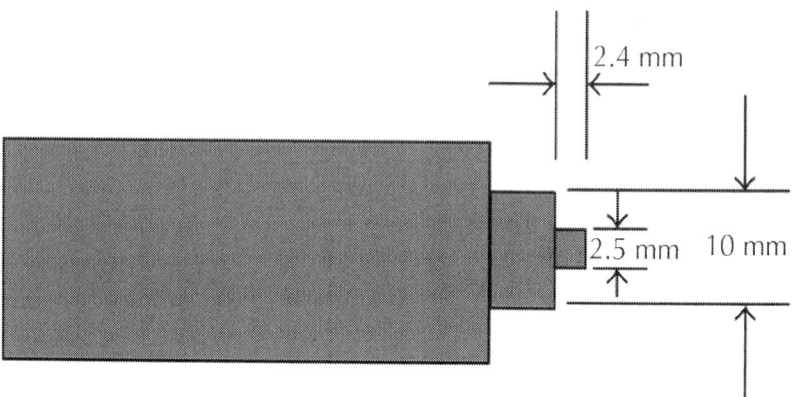

Figure 4: The tool shoulder and pin configuration diagram [8].

In GMAW, the type of the welding current was a DC electropositive current and the shielding gas used was pure argon. The diameter of the aluminum filler wire was 1.0 mm. It had chemical composition of Mn (0.05–0.20%), Si (0.25%), Fe (0.40%), and Cr (05–0.20%) and Al balance with the code reference of ER5356. The heat input was calculated as the ratio of the power (voltage × current) to the velocity of the heat source (the arc) as follows [10]:

$$H = \frac{60EIL}{1000S}, \quad (1)$$

where H, E, and I are the heat input in KJ, the arc voltage in volts, and the current in amps, respectively. S is the travel speed in mm/min. L is the length of weldments in millimeters. The welding current and the welding voltage were measured to be 120 A and 20 V, respectively.

In this investigation, it was found that the travel speed of 45 mm/min for FSW and GMAW is almost similar. The results showed that for FSW sound defect free welds were obtained as opposed to GMAW, where a weld defect was detected.

The heat input per unit length of FSW process was calculated according to the following formula [6]:

$$H = f\sigma_y AL, \tag{2}$$

where f is the ratio in which the heat generated at the tool shoulder/workpiece interface was transported between the tool and the workpiece, σ_y is the yield stress of the metal at $0.8T_s$, T_s is the solidus temperature, and A is the cross-sectional area of the tool shoulder. L is the length of the weldment. The value of j for FSW of aluminum alloy with steel tool is found to be >90%. [6]. Accordingly, the heat input to weld the entire joint of 5 cm in length was calculated for both GMAW and FSW and it was found for GMAW to be four times that of FSW.

Sample preparation for metallography test was performed according to the ASTM E3 (Standard Guide of Metallographic Specimens) [13]. All samples (except for tensile test) were ground, polished, and etched for macrostructure, microstructure, and microhardness test according to the ASTM E407 (Standard Practice for Microetching Metals and Alloys) [14]. The etchant used for aluminum alloys was Keller's reagent which is a mixture of 3 mL hydrochloric acid, 5 mL nitric acid, 2 mL hydrofluoric acid, and 190 mL distilled water.

The tensile test was conducted according to the standard test method for metallic materials (ASTM E8) [15]. The Vickers microhardness test (Hv) was conducted according to the standard test for microhardness of metallic materials (ASTM E384) [16]. The hardness applied load was 1 kg and the dwell time was 8 seconds.

RESULTS AND DISCUSSION

Characterization of the Surface and Root Joints

Figure 5 shows the profile of the Al-Al friction stir welded joints at a rotational speed of 1750 rpm and traverse speed of 45 mm/min.

Figure 5(a) reveals the weld profile at the top of the plate to be rough and spattered. The partially delaminated onion ring pattern is evident. This occurs due to insufficient heat generation in the shoulder rubbing action with the top surface of the plate [17]. In Figure 5(b) the lack of root penetration defect and the incomplete filled groove are eminent. This might be attributed to the shortness of the pin which resulted in poor penetration down to the bottom of the welded piece which in turn resulted in an insufficient heat flow, one that is necessary to deform, plasticize, and join the bottom sides together. The pin plunging depth is a critical factor and difficult to be controlled. The depth of sinking must be constant during the welding process. However, that is very difficult to achieve without assuring a horizontal surface leveling particularly for joining long workpieces [18].

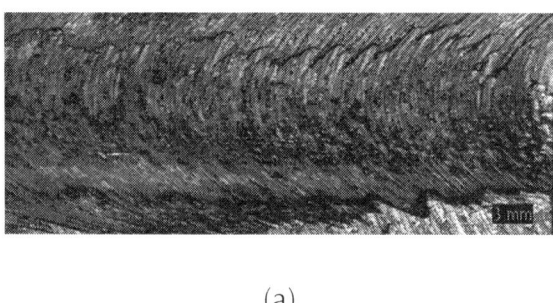

(a)

(b)

Figure 5: The weld profile of the Al-Al friction stir welded joints at a rotational speed of 1750 rpm and traverse speed of 45 mm/min at (a) the top side and (b) the bottom side.

Figure 6 shows the profile of the Al-Al friction stir welded joints at a rotational speed of 2720 rpm. Figure 6(a) depicts much smoother tightly striated weldment profile compared to that joint conducted at a rotational speed of 1750 rpm. Figure 6(b) shows the bottom side of the plate; it reflects a defect-free surface. This was due to sufficient heat received on both top and bottom sides of the plate which enhanced the plasticizing of the aluminium piece during welding. Higher rotational speeds can afford better weld profile even though both the traverse speed and the applied load are kept constant [17].

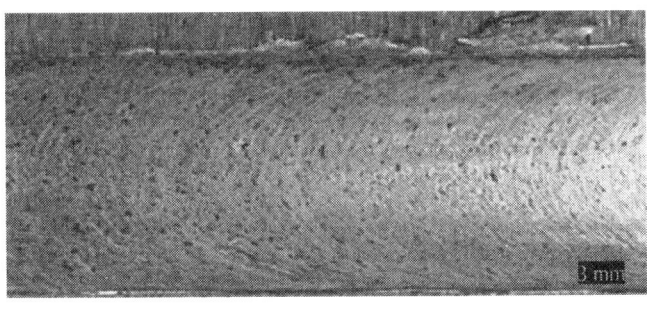

(a)

(b)

Figure 6: The weld profile of friction stir welded Al-Al joint at rotational speed of 2720 rpm and traverse speed of 45 mm/min at (a) top side and (b) bottom side.

Figure 7(a) shows the weld profile of the Al-Al GMAW joint at its top side. Spatters around the welded metals are evident. This might be caused by an excessive current, arc blow, damp electrode, contamination, and/or incorrect wire feed speed during the welding process. Figure 7(b) clearly shows that there is a significance of the lack of penetration occurring at the weld root of the plate. This is probably caused by the high welding speed which results in insufficient heat provisioning and is not enough to melt the bottom of the plate [19].

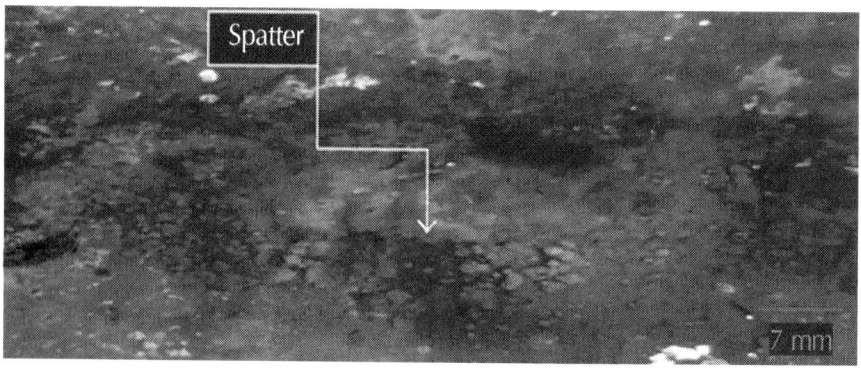

(a)

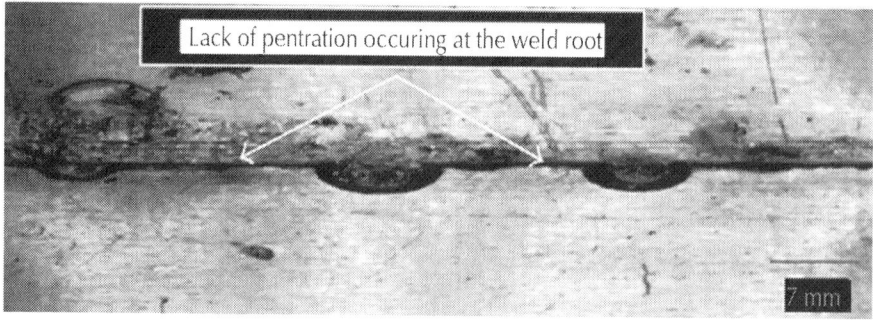

(b)

Figure 7: The weld profile of the Al-Al GMAW welded joint at (a) the top side and (b) the bottom side.

Microstructures Analysis

The optical low magnification image in Figure 8 shows the main features of the FSW process at the NZ. The NZ reflects a basin-like nugget zone shape. The formation of this shape on FSW is attributed to the maximal deformation and plasticization in the material of the upper part of the NZ as opposed to that of its lower parts. The NZ in the pin shoulder side receives the uttermost amount of the frictional heat generated by the intimate contact of its surface with the cylindrical-tool shoulder. Hence it yields, flows, and flattens under the pin swirling action more than its lower part on the anvil side does and results in this basin-like nugget shape. However, so far various shapes of NZ have been observed. The evolution of the NZ shape depends on the processing parameter, tool geometry, and thermal gradient in the workpiece. Mahoney et al. [20] reported elliptical nugget zone shape in the weld of the 7075-T651 aluminum alloy rather than a basin-like shape.

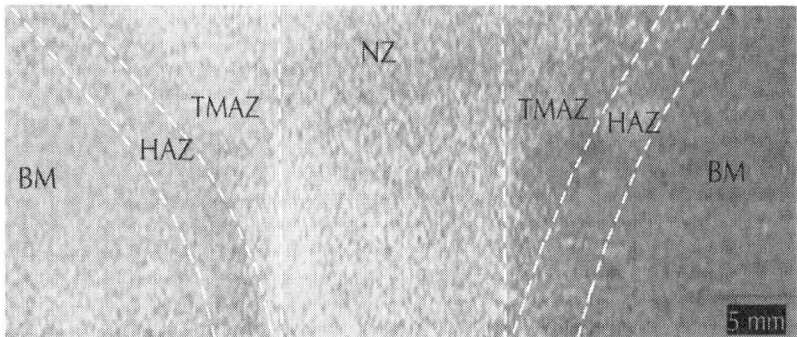

Figure 8: Optical image showing the macroscopic features in a cross section of the FSW butt joint.

Figure 9 shows the microstructure at different welding regions of the Al-Al FSW joints at a rotational speed of 1750 rpm. Figure 10 shows the microstructure at different welding regions of the Al-Al FSW joints with a rotational speed of 2720 rpm. Figures 9(a) and 10(a) show the microstructure of the original as-received base metal. The randomly distributed second phase particles appeared as small black particles. Elongation of the Al grains along the rolling direction is evident. During FSW, the base metal experiences no metallurgical changes and maintains its original cold-worked microstructure. Figures 9(b) and 10(b) show that the two NZs of the weld joints resulted from the two different rotational speeds. NZ produced at 2720 rpm has a smaller grain size as opposed to that conducted at a speed of 1750 rpm. These variations result from excess heat exposure and the intensive stirring action induced at higher rotational speeds.

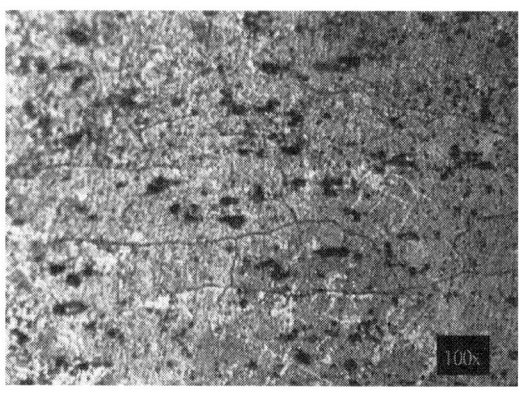

(a)

(b)

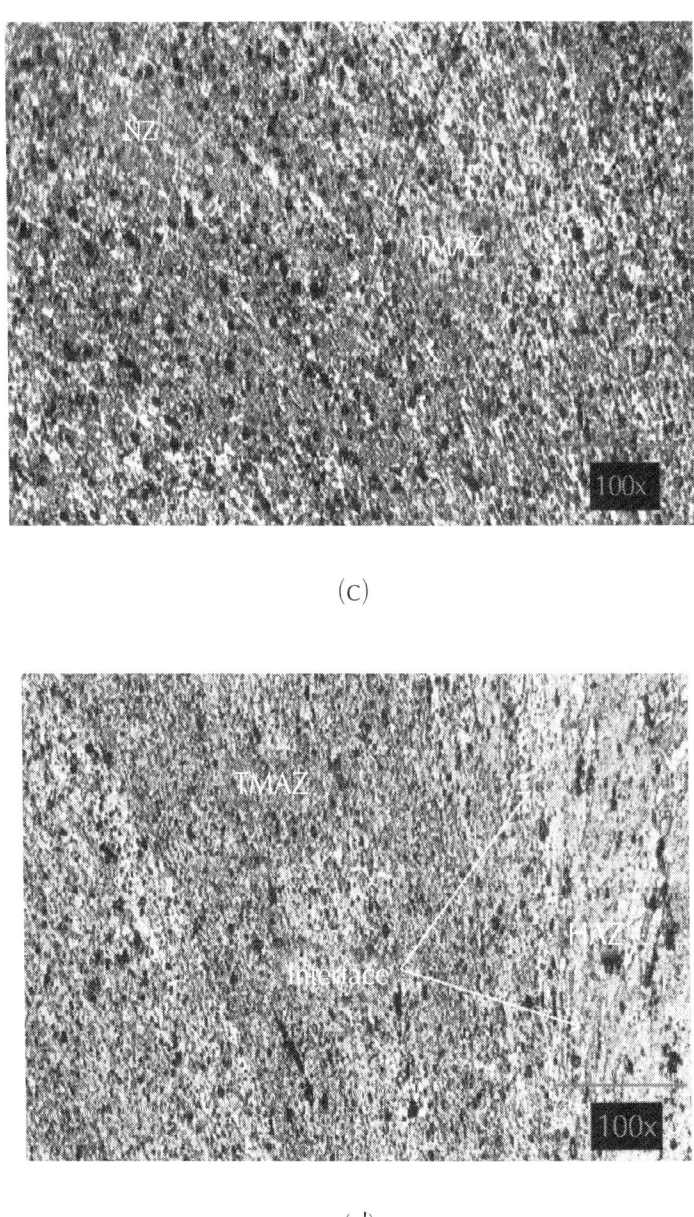

(c)

(d)

Figure 9: Optical images reveal the microstructures of different welding zones on Al-Al butt welded joint performed by the FSW technique at a rotational speed of 1750 rpm. (a) BM, (b) NZ, (c) NZ and TMAZ, and (d) TMAZ and HAZ 100x magnification.

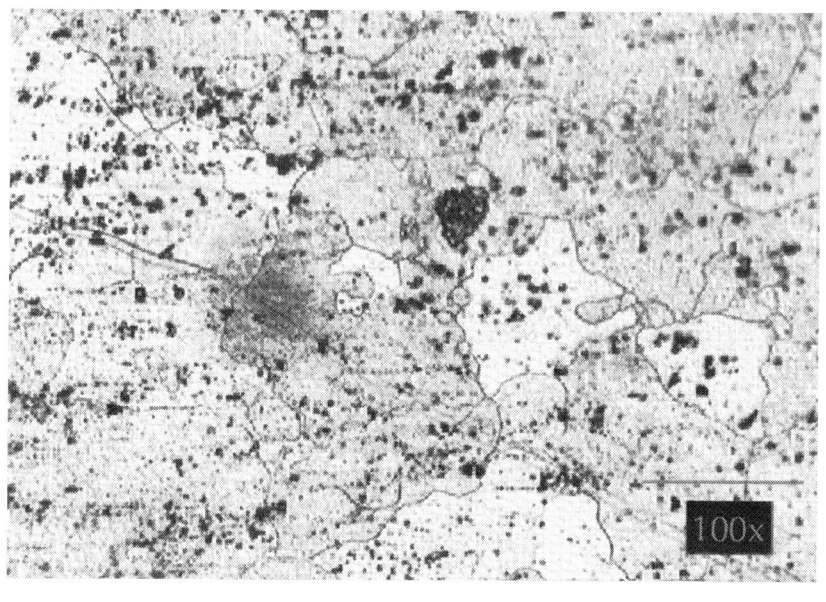

(a)

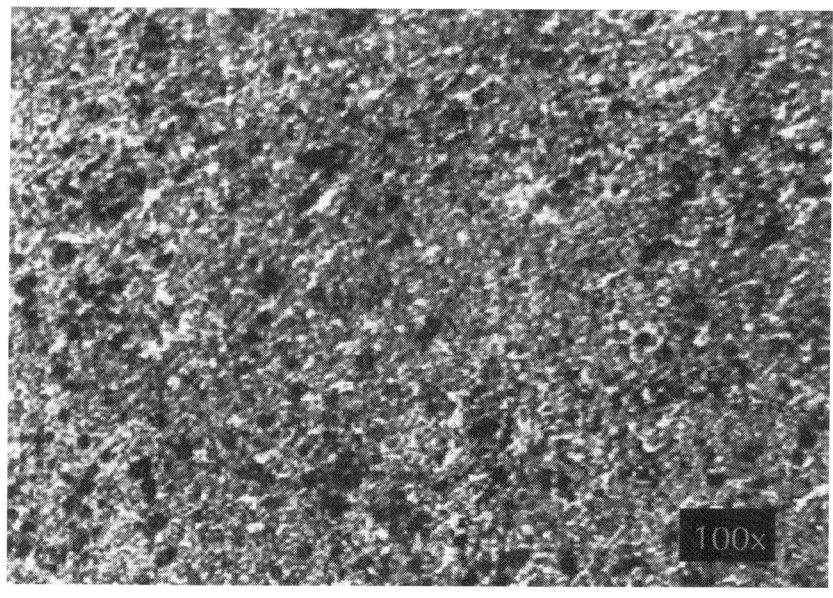

(b)

Advantages of the Green Solid State FSW over the ... 107

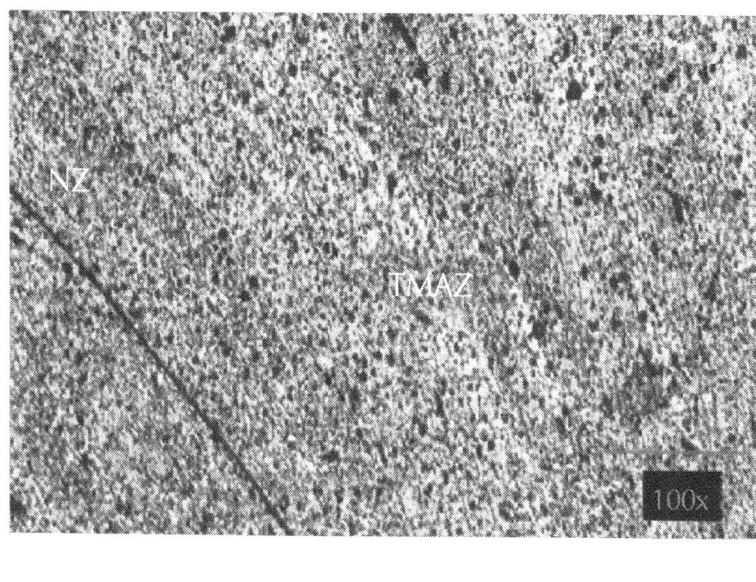

(c)

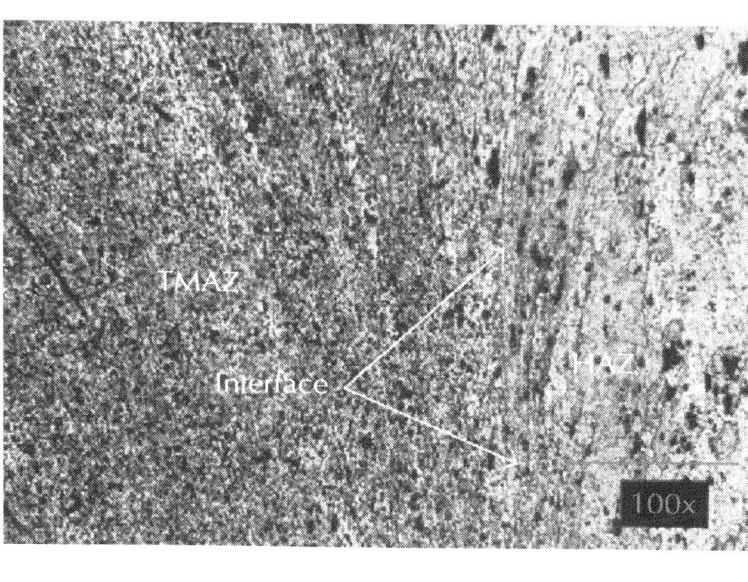

(d)

Figure 10: Optical images reveal the microstructures of different welding zones on Al-Al butt welded joint performed by FSW at a rotational speed of 2720 rpm. (a) BM, (b) NZ, (c) NZ and TMAZ, and (d) TMAZ and HAZ 100x magnification.

Figures 9(c) and 10(c) show the NZ and the TMAZ at the two rotational speeds. It is prominent that the grain size at TMAZ is larger than the grain size at NZ. Regarding the grain size phylogeny at the NZ and the TMAZ, there are two conflicting factors acting together and affecting this process. During welding, the grains at these two regions undergo plastic deformations, mechanical shearing, and thermal exposure. Hence, they get plasticized, extruded, and rotated according to the strain levels to which they are subjected [21]. The mechanical actions of deformation, shearing, and extrusion result in smaller grain size, while the thermal exposure acts as grain size promoter. In the NZ, the mechanical actions are more severe and the outcome of these conflicting actions work in favor of dynamic recrystallization resulting in small grain size, smaller than that at the BM, TMAZ, and HAZ. This mechanical action is augmented at higher speeds and results in even smaller grain size. The TMAZ is located slightly further away from the direct effect of the rotating pin action. It is characterized by a highly deformed structure. Although the TMAZ undergoes plastic deformation, but dynamic crystallization does not occur in this zone due to insufficient deformation and strain. Consequently the thermal and mechanical interactions work in favor of thermal effect and result in grain growth. Higher rotational speed favors larger grain size. The grain size decreases as the input heat decreases.

Figures 9(d) and 10(d) show the TMAZ and the HAZ of the Al-Al welded joints. The grains at the HAZ are coarser as opposed to those at the TMAZ. The thermal exposure causes the grains to grow. The grains become coarser as they are located further away from the center of the NZ.

The grain size at different welding regions for the two FSW Al-Al joints conducted at 1750 and 2720 rpm is shown in Table 1. The relation between grain size and mechanical properties can be expressed by the following Hall-Petch equation [22]:

$$\sigma = \sigma_0 + K_h d^{-1/2}, \qquad (3)$$

Where σ is the strength of the material, d is the grain size diameter, σ_0 and K_h are experimental constants and are different for each metal. Equation (3) shows that smaller grain size diameter results in higher microhardness and UTS of the material. Besides, the mechanical properties not only depend on the degree of grain refinement. And in addition, secondary phase formation, microstructure homogeneity, and microstructure defects play a vital role in deciding the mechanical properties of the weld joint.

Table 1: Grain size at different welding zones for FSW and GMAW for Al-Al joints

Items	Range of grain size in μ m of the weld regions			
	NZ/WZ	TMAZ	HAZ	BM
FSW				
Rotational speed				
1750 rpm	3.5–10	3–11	8–18	5–16
2720 rpm	2–8	3.5–13	8–19	5–16
GMAW	7–12	—	10–22	5–16

For GMAW, the solidified weld pool and the corresponding heat affected base metal are called the welding zone (WZ) and the heat affected zone (HAZ), respectively. Partially melted zone (PMZ) in the interface of WZ and HAZ is also found in the case of some specific nonferrous alloys. The width of these zones is a direct function of the input heat and material's thermal conductivity. The cooling rate and peak temperature primarily dictate the solidification mode and phase content of the weld microstructure. The initial grain morphology is found to be columnar dendrites and develop with a different inclination to equiaxed dendritic [23]. Figure 11 shows the microstructures of (a) BM and (b) WZ and HAZ of a weld joint produced by GMAW similar to that produced by FSW. Evolution of dendritic structure in WZ is attributed to the fast heating of the weld zone up to the melting temperature and the subsequent fast cooling of the molten pool [24]. However, FSW requires much lower heat input and welding power compared to the arc welding process.

Less heat input helps improve the joint mechanical properties and decreases both distortion and residual stresses [19].

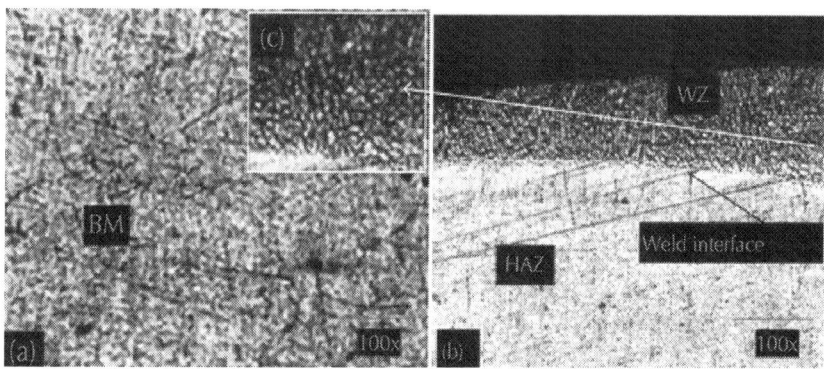

Figure 11: The welding morphology of the Al-Al GMAW joint depicts the microstructure of the (a) BM and (b) WZ and HAZ. The transition of grain structure from columnar dendritic to equiaxed dendritic is illustrated in the enlarged region of WZ in (c).

Environmental Effects of FSW and GMAW Techniques

Indoor Air Quality Pro device was used to detect and analyze the amount of emitted gases during the welding process. The amounts of the detected gases for both FSW and GMAW processes were compared to determine which welding technique results in the release of more harmful gases to the surroundings. The measured amount of carbon dioxide and carbon monoxide gases prior to welding and after welding is shown in Table 2. To assure reliable results, the measurements were conducted in a closed confined volume of 7.2 m^3 welding stalls.

Table 2: The amounts of the released gases in the welding area prior to and after welding

Number of tests	Atmosphere		GMAW		FSW	
	Carbon monoxide [ppm]	Carbon dioxide [ppm]	Carbon monoxide [ppm]	Carbon dioxide [ppm]	Carbon monoxide [ppm]	Carbon dioxide [ppm]
1	0.4	121	2.8	361	0.5	197
2	0.3	118	2.0	354	0.3	241
3	0.5	122	2.2	344	0.6	196
4	0.7	119	2.7	338	0.7	201
5	0.9	117	3.9	333	1.0	223

Scanning Electron Microscopy (SEM) Analysis

Figure 12 shows an SEM image coupled with EDX scan analysis of the NZ of Al-Al FSW joint produced at a rotational speed of 2720 rpm. Weak peaks of carbon and oxygen can be seen in the EDX plot analysis; oxidization of the Al-Al welded joints at the NZ is expectable. High frictional heat generation and atmosphere humidity of 75% enhance oxidation of aluminum during the welding process. Existence of carbon in the NZ may be attributed to the contamination from the medium carbon steel tool during the welding process. Surprisingly Fe was not detected by the EDX analysis. However, its existence within the NZ is confirmed by the X-ray diffraction (XRD) analysis.

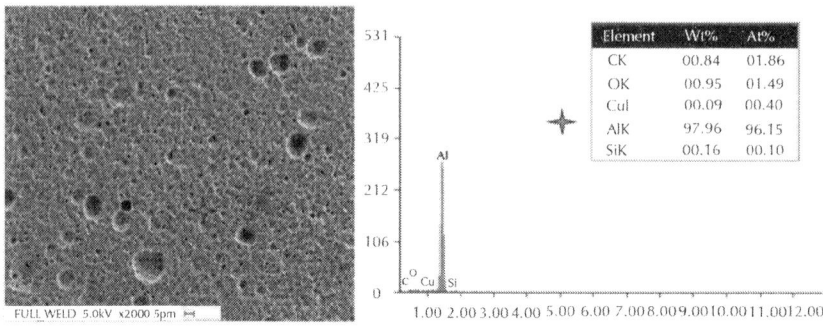

Figure 12: SEM image and EDX analysis of the NZ of Al-Al FSW joint performed at a rotational speed of 2720 rpm.

X-Ray Diffraction (XRD) Analysis

XRD analysis was performed on the base metal and the NZ for the two rotating speeds. The resulting superimposed diffraction patterns are displayed in Figure 13. Besides, α Al existence of different intermetallic phases at the NZ of the two rotating speeds is evident. These intermetallic phases often have complex lattice structures and microhardness values. The Al-Fe-Si ternary system is the key for Al alloy family phase formation. The formation of different Al-Fe and Al-Fe-Si intermetallic phases at different rotating speeds is attributed to the amount of heat input and the Fe content in the nugget zone. Stirring action at different rotating speeds leads to transferring of different Fe amounts from the steel tool to the welded piece and eventually results in the formation of different intermetallic phases at various temperatures. Despite the high hardness of these intermetallic phases, they have some negative effects, especially that of Al-Fe intermetallic such as $Al_{13}Fe_4$ and $AlFe_3$. The large electrochemical potential of 1.22 volts between iron and aluminum results in higher susceptibility to intercrystalline and galvanic corrosion [25].

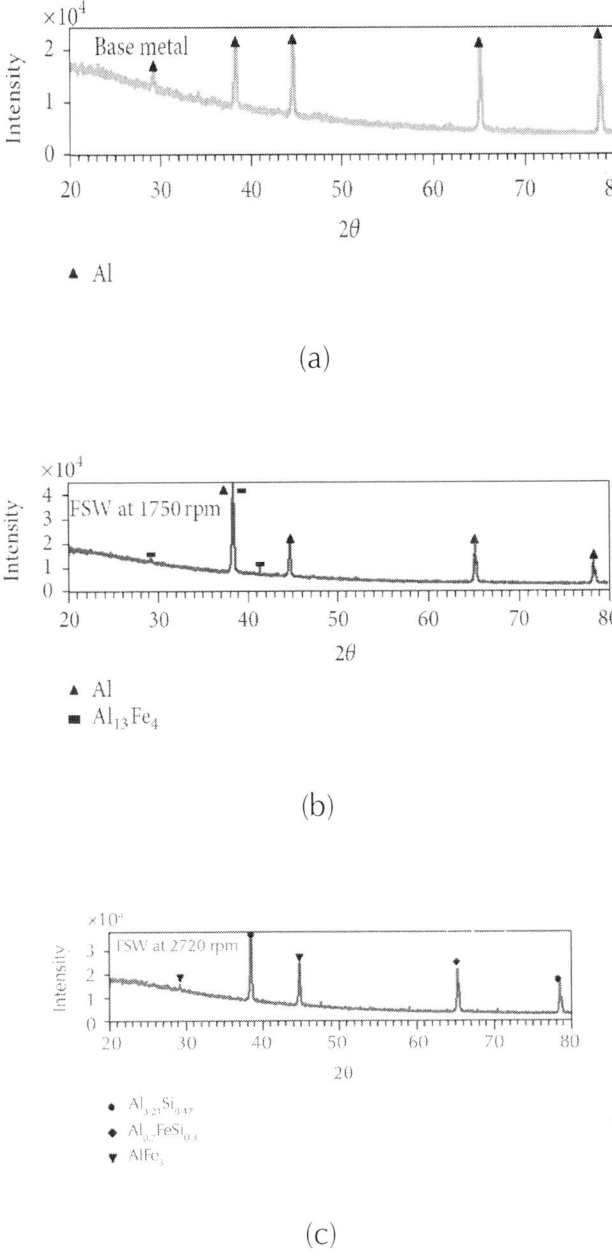

Figure 13: X-ray superimposed diffraction patterns of the Al-Al FSW joint. (a) BM and (b) NZ at rotating speed of 1750 rpm and (c) NZ at rotating speed of 2720 rpm.

X-ray diffraction analysis was performed also to identify the phases of the Al-Al GMAW joints. According to the XRD analysis the usage of the filler wire does not affect on the phase transformation. This may be attributed to the difference in phase formation mechanisms between FSW and GMAW processes. The XRD pattern in Figure 14 reveals that mainly pure Al phase is present.

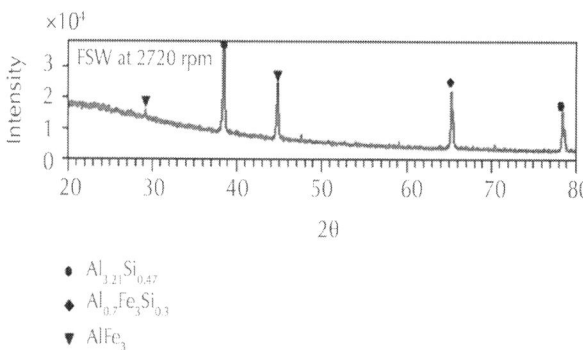

- $Al_{3.21}Si_{0.47}$
- $Al_{0.7}Fe_3Si_{0.3}$
- $AlFe_3$

Figure 14: X-ray diffraction pattern of the Al-Al GMAW joint. Mainly peaks of pure are eminent.

Tensile Test

Figure 15 clearly illustrates the inferior strength of all weld joints as opposed to the UTS of the base metal. Among the three fabricated weld joint sets, the Al-Al FSW joints conducted at a speed of 2720 rpm gave the highest tensile strength. This was due to the grain refinement; homogeneous and defect-free microstructure of the joint resulted from the very intense stirring action at high rotational speed. The Al-Al FSW joints conducted at a speed of 1750 rpm have slightly inferior tensile strength as compared to that conducted at 2720 rpm speed. The tensile tests demonstrated that fractures mainly occurred at the boundary between the nugget and the TMAZ rather than along interface within the nugget itself.

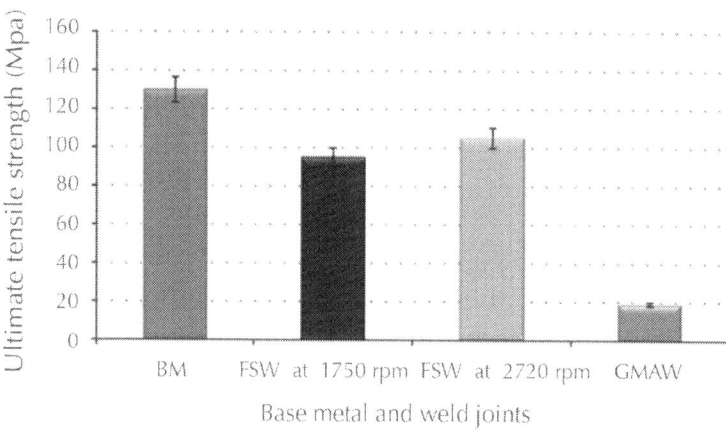

Figure 15: The UTS of the base metal and the weld joints produced by the FSW and the GMAW techniques.

The Al-Al GMAW joint possesses the low tensile strength as compared to the other fabricated weld joints. The reduction in the UTS of this weld was about 85.7%. It only possessed about 19.7% and 17.7% of the UTS of the welded joints fabricated by the FSW technique at the two rotation speeds of 1750 rpm and 2720 rpm, respectively. This high reduction in the tensile strength experienced by the GMAW joints thought to be due to the evolution of many defects during the melting and solidification process such as impurities, cracks, and pores adversely affects mechanical properties.

Microhardness Test

Figure 16 depicted that the microhardness at the NZ, TMAZ, and HAZ for the FSW Al-Al joints is lower than the microhardness of the BM. This is attributed to the effects of thermal exposure and mechanical action which lead to different levels of dynamic recrystallization, grain growth, and phase transformation at the different welding zones. Recrystallization occurs when new grains form and consume the original cold-worked grains. The higher microhardness and strength of the BM are ascribed to the cold working gained during the production stages of casting, rolling, and forming. Cold working enhances microhardness and strength. It causes dislocations to be entangled with one another and hinders their motion [26].

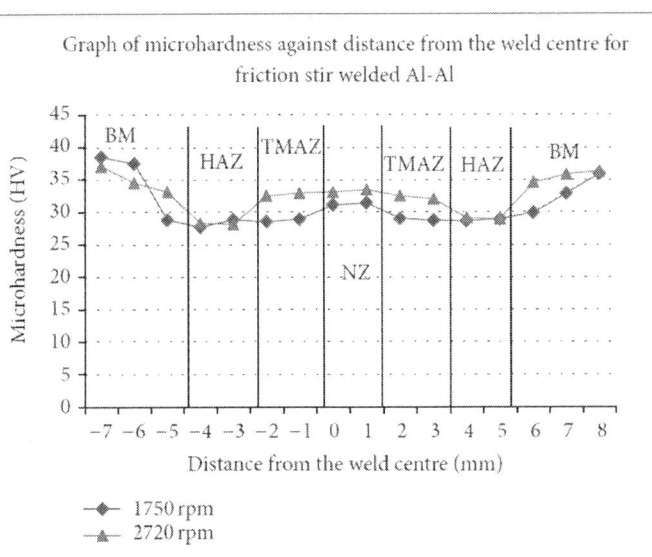

Figure 16: The microhardness profiles of Al-Al joints fabricated by FSW.

The NZ showed the highest microhardness values as compared to other weld zones. HAZ showed the lowest hardness values. In fact, this is understandable. The HAZ is far away from the stirring action influence. This justifies why the microhardness values of the HAZ at the two rotational speeds were almost identical.

Figure 17 exhibits the microhardness profile for the Al-Al GMAW joints. Again the microhardness values of the weld zones are much lower than that of the BM. The elongated dendritic and the equiaxed grain structure with the WZ result in the fast heating and subsequent fast cooling of the molten pool. The bulk volume of the workpiece acts as an efficient sink for the heat generated during welding and results in different cooling rates at the HAZ and the WZ [27]. The heat flow gradient affects the mechanical properties of the joint. The HAZ demonstrated lower hardness than that of the WZ specifically near the weld interface. The weld interface is a narrow boundary that separates the WZ and the HAZ. It consists of a thin band of the BM that has melted or partially melted during the welding process but has immediately solidified before any mixing could take place.

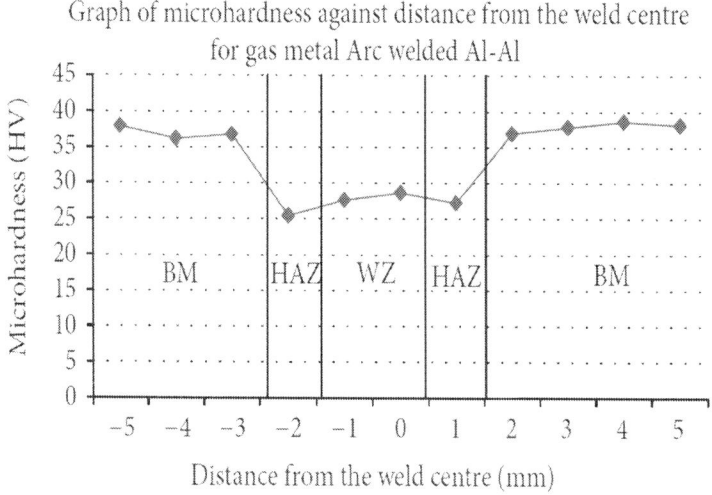

Figure 17: The microhardness profiles of Al-Al joint fabricated by GMAW.

CONCLUSIONS

- In this study FSW technique showed outstanding advantages over the GMAW process in joining 1030 Al alloys. It gave clean, defect-free, and better weldment microstructure which in turn resulted in superior mechanical properties.
- FSW showed a substantial saving in power consumption. The power consumed in GMAW was fourfold that of FSW for execution of similar joints. This resulted in a reduction in the area where microhardness changes. The HAZ in FSW was narrower than that in GMAW process.

- GMAW process released higher amounts of harmful gases such as carbon monoxide and carbon dioxide to the surroundings (2.7 ppm and 346 ppm, resp.) as opposed to 0.6 ppm and 211.6 ppm, respectively, for FSW.
- The joints welded by GMAW process exhibited substantial reduction in their UTS and microhardness values, 80% and 12%, respectively, as opposed to FSW joints executed at rotation speed of 2720 rpm, 78% and 9.5%, respectively, as opposed to FSW joints executed at 1750 rpm.

ACKNOWLEDGMENTS

This work is supported under the Grant no. 9001-00338 of the Universiti Malaysia Perlis (UniMAP). The authors gratefully acknowledge the outstanding support provided by the technicians of the workshop in the Materials Engineering School, UniMAP.

REFERENCES

1. R. K. Shukla and P. K. Shah, "Comparative study of friction stir welding and tungsten inert gas welding process," Indian Journal of Science and Technology, vol. 3, no. 6, pp. 667–671, 2010.
2. E. D. Nicholas, "Friction stir welding-a decade on," in Proceedings of the IIW Asian Pacific International Congress, Sydney, Australia, October-November 2000.
3. O. Vladvoj, S. Margarita, F. D. S. Jorge, and V. Pedro, "Microstructure and properties of friction stir welded aluminium alloys," in Paper Presented at Metal, Hradec nad Moravicí, Czech Republic, 2005.
4. A. V. Kumar and K. Balachandar, "Effect of welding parameters on metallurgical properties of friction stir welded aluminium alloy 6063-O," Journal of Applied Sciences, vol. 12, no. 12, pp. 1255–1264, 2012.

5. W. H. Jiang and R. Kovacevic, "Feasibility study of friction stir welding of 6061-T6 aluminium alloy with AISI 1018 steel," Proceedings of the Institution of Mechanical Engineers B: Journal of Engineering Manufacture, vol. 218, no. 10, pp. 1323–1331, 2004.
6. G. G. Roy, R. Nandan, and T. DebRoy, "Dimensionless correlation to estimate peak temperature during friction stir welding," Science and Technology of Welding & Joining, vol. 11, no. 5, pp. 606–608, 2006.
7. T. Hirata, T. Oguri, H. Hagino, et al., "Influence of friction stir welding parameters on grain size and formability in 5083 aluminum alloy," Materials Science and Engineering A, vol. 456, no. 1, pp. 344–349, 2007.
8. J. N. Pires, A. Loureiro, and G. Bölmsjo, Welding Robots, Springer, London, UK, 2006.
9. R. W. Messler, Principles of Welding: Processes, Physics, Chemistry, and Metallurgy, John Wiley & Sons, New York, NY, USA, 2008.
10. S. Kou, Welding Metallurgy, Wiley-Interscience, Hoboken, NJ, USA, 2nd edition, 2003.
11. M. Esmaily, S. N. Mortazavi, P. Todehfalah, and M. Rashidi, "Microstructural characterization and formation of martensite phase in Ti-6Al-4V alloy butt joints produced by friction stir and gas tungsten arc welding processes," Materials and Design, vol. 47, pp. 143–150, 2013.
12. P. B. Anjaneya and P. Prasanna, "Experimental comparison of the MIG and friction stir welding processes for AA 6061(Al Mg Si Cu) aluminium alloy," International Journal of Mining, Metallurgy & Mechanical Engineering, vol. 1, no. 2, pp. 137–140, 2013.
13. ASTM Standard E3, "Standard Guide for Preparation of Metallographic Specimens".
14. ASTM Standard E407, "Standard Practice for Microetching Metals and Alloys".

15. ASTM Standard E8, "Standard Test Methods for Tension Testing of Metallic Materials".
16. ASTM Standard E384, "Standard Test Method for Knoop and Vickers Hardness of Materials".
17. P. Bahemmat, A. Rahbari, M. Haghpanahi, and M. K. Besharati, "Experimental study on the effect of rotational speed and tool pin profile on AA2024 aluminum friction stir welded butt joints," pp. 1.1–1.7, Proceedings of the ASME Early Career Technical Conference (ECTC ‹08), Miami, Fla, USA, October 2008.
18. C. Meran, "The joint properties of brass plates by friction stir welding," Materials & Design, vol. 27, no. 9, pp. 719–726, 2006.
19. S. R. McLaughlin, C. J. Bayley, and N. M. Aucoin, "Assessment of microstructural heterogeneities in multipass pulsed gas metal arc welds," Canadian Metallurgical Quarterly, vol. 51, no. 3, pp. 294–301, 2012.
20. M. W. Mahoney, C. G. Rhodes, J. G. Flintoff, W. H. Bingel, and R. A. Spurling, "Properties of friction-stir-welded 7075 T651 aluminum," Metallurgical and Materials Transactions A, vol. 29, no. 7, pp. 1955–1964, 1998.
21. G. Liu, L. E. Murr, C.-S. Niou, J. C. McClure, and F. R. Vega, "Microstructural aspects of the friction-stir welding of 6061-T6 aluminum," Scripta Materialia, vol. 37, no. 3, pp. 355–361, 1997.
22. J. W. Dini, Electrodeposition: The Materials Science of Coatings and Substrates, Noyes Publications, Park Ridge, NJ, USA, 1993.
23. P. Kamal and K. P. Surjya, "Effect of pulse parameters on weld quality in pulsed gas metal arc welding: a review," Journal of Materials Engineering and Performance, vol. 20, no. 6, pp. 918–931, 2011.
24. Y. Liu, W. Wang, J. Xie, et al., "Microstructure and mechanical properties of aluminum 5083 weldments by gas tungsten arc

and gas metal arc welding," Materials Science and Engineering A, vol. 549, pp. 7–13, 2012.

25. P. B. Srinivasan, K. S. Arora, W. Dietzel, S. Pandey, and M. K. Schaper, "Characterisation of microstructure, mechanical properties and corrosion behaviour of an AA2219 friction stir weldment,"Journal of Alloys and Compounds, vol. 492, no. 1-2, pp. 631–637, 2010.

26. H. J. McQueen, "Recovery and recrystallization during high temperature deformation," Treatise on Materials Science & Technology, vol. 6, pp. 393–493, 1975.

27. A. K. Lakshminarayanan, V. Balasubramanian, and K. Elangovan, "Effect of welding processes on tensile properties of AA6061 aluminium alloy joints," The International Journal of Advanced Manufacturing Technology, vol. 40, no. 3-4, pp. 286–296, 2009.

Corrosion Behavior of Carbon Steel in Synthetically Produced Oil Field Seawater

Subir Paul, Anjan Pattanayak, and Sujit K. Guchhait

Department of Metallurgical and Material Engineering, Jadavpur University, Kolkata 700032, India

ABSTRACT

The life of offshore steel structure in the oil production units is decided by the huge corrosive degradation due to, SO_4^{2-}, S^{2-} and Cl^-, which normally present in the oil field seawater. Variation in pH and temperature further adds to the rate of degradation on steel. Corrosion behavior of mild steel is investigated through polarization, EIS, XRD, and optical and SEM microscopy. The effect of all 3 species is huge material degradation with FeS_x and $FeCl_3$ and their complex as corrosion products. EIS data match the model of Randle circuit with Warburg resistance. Addition of more corrosion species decreases impedance and increases capacitance values of the Randle circuit at the interface. The attack is found to be at the grain boundary as well as grain body with very prominent sulphide corrosion crack.

INTRODUCTION

The severe corrosion of the submersed structures in the oil field at the production site and crude oil transportation is unpredictable and is a major component of the total corrosion loss in oil and gas industries. The corrosion species in the aqueous oil field seawater are CO_3^{2-}, S^{2-}, Cl^-, SO_4^{2-}, and O (Table 1) [1] which is also influenced by the variation of pH and temperature. CO_3^{2-} and S^{2-} are formed from CO_2 and H_2S of the oil in the aqueous environment. And Cl^-, SO_4^{2-}, and O are present in the seawater. Besides these parameters, there are fluid dynamics of sea water and suspended solids and sands, influencing the erosion corrosion of the marine structures. Crude oil and natural gas can carry various high-impurity products which are inherently corrosive. In the case of oil and gas wells and pipelines, such highly corrosive media are carbon dioxide (CO_2), hydrogen sulfide (H_2S), and free water [2].

Table 1: Ions present in typical oil field seawater

Species typically found in oil field brines		Element		Concentration mg/L
CO_2	Dissolved carbon dioxide	Barium	Ba^{2+}	31
H_2CO_3	Carbonic acid	Boron	B	6

HCO_3^-	Bicarbonate ion	Calcium	Ca^{2+}	284
CO_3^{2-}	Carbonate ion	Iron	Fe^{3+}	55.85
H^+	Hydrogen ion	Magnesium	Mg^{2+}	24.31
OH^-	Hydroxide ion	Phosphorous	P^{3-}	1
Fe^{2+}	Iron ion	Potassium	K	50
Cl^-	Chloride ion	Sodium	Na	4770
Na^+	Sodium ion	Strontium	Sr^{2+}	83
K^+	Potassium	Chloride	Cl^-	7480
Ca^{2+}	Calcium ion	Bromide	Br^-	20
Mg^{2+}	Magnesium ion	Sulphate	SO_4^{2-}	21
Ba^{2+}	Barium ion	Nitrate	NO_3^-	0.50
Sr^{2+}	Strontium ion	Hydroxyl	OH^-	0
CH_3COOH (HAc)	Acetic acid	Carbonate	CO_3^{2-}	0
CH_3COO^- (Ac^-)	Acetate ion	Bicarbonate	HCO_3^-	500
HSO_4^-	Bisulphate ion	Dissolved CO_2	CO_2	92.4
SO_4^{2-}	Sulphate ion	Specific gravity		1.014
		pH		6.58
		Resistivity		0.4405 Ohm
		Total dissolved solids 1		3.453 mg

The effect of any individual parameter on corrosion rate has been studied extensively [3–6]. But the conjoint effect of the above mentioned parameters and interfering effects and interactions are complex and are not very well understood. The salts and sulfide compounds dissolved in crude oil can provoke the formation of a corrosive aqueous solution whose chemical composition involves the presence of both hydrochloric acid (HCl) and hydrogen sulfide (H_2S) [3, 4]. Corrosion mitigation in the oil field industry has traditionally been performed by combining methods for measuring the corrosion rates such as corrosion coupons and regular pipeline inspections with prevention strategies [6]. But that required years to get empirical results and could not be applied to other geographical locations of different sea water chemistry. All the factors make the corrosion mechanisms in the oil fields very complex with high degree of interaction among the species. Several previous studies have been performed related to the corrosion process of iron and steel in H_2S solutions [4, 7–13]. These works studied the influence of H_2S on the corrosion phenomena at ambient temperature. In H_2S-containing solutions, the corrosion process of metal may be accompanied by the formation of a sulfide film on the metal surface and leads to more complicated corrosion behavior. Previous researches [14–16] have shown that H_2S had a remarkable acceleration effect on both the anodic iron dissolution and the cathodic evolution in most cases but H_2S may exhibit an inhibitive effect on the corrosion of iron or steel weld. Recently, the influence of H_2S concentration on the corrosion behavior of carbon steel at 90°C has been investigated [15]. Physical modeling of ships and offshore structures in ocean water by Melchers et al. [17–19] and Shehadeh and Hassan [20] adds to better understanding of the present investigation. However, little research has been done on the corrosion behavior of carbon steel in the presence of both H_2S and NaCl at ambient and elevated temperature.

Corrosion mechanisms in oil field systems are complex and are showing high degrees of interaction between corrosion species, products, and oil field metallurgies. The interactions of sulfate and chloride are of interest in this work, since presence of sulfate ions, in oilfield produced water, strongly influence corrosion mechanisms. While there are many research works on the effects of CO_2 and H_2S on corrosion of carbon steel, those of conjoint effects of S^{2-}, Cl^-, and SO_4^{2-} are much less. The present investigation aims to study the conjoint effects of S^{2-}, Cl^-, and SO_4^{2-} along with variation of pH and temperature on carbon steel. The corrosive species included are sulfate, chloride, hydrogen sulfide, temperature, and pH. Sulphate and chloride were added as Na_2SO_4 and NaCl. Hydrogen sulfide was introduced to the corrosion cell with the following reaction:

$$Na_2S + H_2SO_4 = H_2S + FeSO_4 \tag{1}$$

The effect on corrosion of these species was examined through polarization experimentation using a three-electrode glass corrosion cell and potentiostat. Electrochemical AC impedance spectroscopy studies were also carried out for better understanding of electrochemical effects of corrosive species on electrical phenomenon occurring at metal-solution interface. The corroded and uncorroded substrates were characterized by XRD. The morphology of the corroded surface was investigated by optical microscopy and SEM.

EXPERIMENTAL METHODS

Polarization Studies

Electrochemical measurements were conducted using Gamry Potentiostat instrument coupled with Echem analyst software, controlled by a personal computer, in a conventional three-electrode cell systems. The working electrode was carbon steel, the counter electrode was graphite, and saturated calomel electrode (SCE) acted as the reference electrode. Experiments were performed in different concentrations of Cl^-, SO_4^{2-}, and S^{2-} solutions, at preselected pH and temperature, to determine the corrosion potential E_{corr} and corrosion current i_{corr}. The potential was scanned between -1.5 V and 1 V at a scan rate of 1 mV/s.

Electrochemical Impedance Spectroscopy (EIS)

The experimental arrangement was the same as that of polarization studies. The electrochemical cell was connected to an impedance analyzer (EIS300 controlled by Echem analyst software) for electrochemical impedance spectroscopy. The electrochemical impedance spectra were obtained at frequencies between 300 kHz and 0.01 Hz. The amplitude of the sinusoidal wave was 10 mV.

The following results and information are obtained from the EIS experiments: Polarization resistance (R_p), electrolyte resistance (R_u), double layer capacitance (C_{dl}), capacitive load or constant phase element, CPE(Y), and α which is defined from the capacitive impedance equation $.Z=1/C(jw)^{-a}$

Capacitors in EIS experiments often do not behave ideally. Instead, they act like a constant phase element (CPE). The exponent α=1 for pure capacitance. For a constant phase element, the exponent α is less than one. The "double layer capacitor" on real cells often behaves like a CPE instead of like a pure capacitor.

X-Ray Diffraction (XRD) Analysis

The X-ray diffraction technique is used to define the crystalline structure and the crystalline phases. This test was done using a Rigaku Ultima III X-Ray Diffractometer for recording the diffraction traces of the samples with monochromatized Cu $K_α$ radiation, at room temperature; the scan region (2θ) ranged from 10° to 100° at a scan rate of 5° min^{-1}.

Scanning Electron Microscope (SEM) Morphology

The electron micrographs were studied by SEM with accelerating voltage 30 kV, magnification up to 300,000x, and resolution of 3.5 nm. The images of the corroded samples were photographed at low and high magnification.

RESULTS AND DISCUSSIONS

The effects of Cl^-, SO_4^{2-}, S^{2-}, pH, and temperature on degradation behavior of carbon steel were studied by potentiostatic polarization to determine corrosion current and corrosion potential. The various electrical properties at the metal-solution interface were determined by electrochemical impedance spectroscopy (EIS). The presence of different elements on corroded surface was detected by XRD. The morphology of the degraded surfaces was characterized by optical microscopy and SEM. Before going into the experimental findings of the effects of different interfering ions, it is worthwhile to discuss the basic electrochemical reactions of aqueous corrosion of steel in the presence of those ions.

The main electrochemical anodic and cathodic reactions for the corrosion of carbon steel in aqueous oil fields environments in presences of the ions are as follows.

Half Cell Reactions (E versus SCE) Consider

$$Fe = Fe^{2+} + 2e \quad (E^0 = -.681) \tag{2}$$

$$O_2 + 2H_2O + 4e^- \longrightarrow 4OH^- \quad (E^0 = 0.579) \tag{3}$$

$$H + e = \frac{1}{2}H_2 \quad (E^0 = -0.241) \tag{4}$$

In presence of SO_4^{2-}, present reactions are

$$SO_4^{2-} + 2e + H_2O = 2SO_3^{2-} + OH \quad (E^0 = -1.177) \tag{5}$$

$$2SO_3^{2-} + 4e + 3H_2O = 3S_2O_3^{2-} + 6OH^- \quad (E^0 = -0.99) \tag{6}$$

$$2SO_3^{2-} + 4e + 3H_2O = S + 6OH^- \tag{7}$$

$$S + 2e^- = S^{2-} \quad (E^0 = -0.688) \tag{8}$$

In presence of Na_2S or H_2S (Na_2S was added in state of H_2S to understand the effect of s^{2-}),

$$H_2S = H^+ + HS^- \tag{9}$$

$$HS^- = H^+ + S^- \quad (E^0 = -0.688) \tag{10}$$

In the presence of Cl^-, that is (NaCl/HCl),

$$Fe^{2+} + 2Cl^- = FeCl_2 \tag{11}$$

$$FeCl_2 + 2H_2O = Fe(OH)_2 + 2H^+ + 2Cl^- \tag{12}$$

$$FeCl_2 + Cl^- = FeCl_3 + e \tag{13}$$

$$Fe + 2H^+ = Fe^{2+} \tag{14}$$

The above equations would help in better understanding of the effects of the corrosion species found in the experimental results.

Polarization Studies

Effect of SO_4^{2-}

Figure 1 shows the potential dynamic polarization curve with increasing SO_4^{2-} concentration at pH 6. The pH of sea water normally varies from 7.5 to 8.2. But in the oil fields due to the presence of few acidic substances, namely, carbonic acid, H_2S, and other organic acids, pH may shift from near neutral towards the acidic side between 6 and 4. It is seen here that the corrosion rate increases with increase in concentration of SO_4^{2-}. It is seen from (5)–(8) that the cathodic reduction of SO_4^{2-} ions produces thiosulfate and sulphide ions, both of which are aggressive corrosion species and hence degrade the steel surface in sea water in the acidic pH range below the neutral medium. The corrosion product in this case should be iron sulphide or thiosulfate.

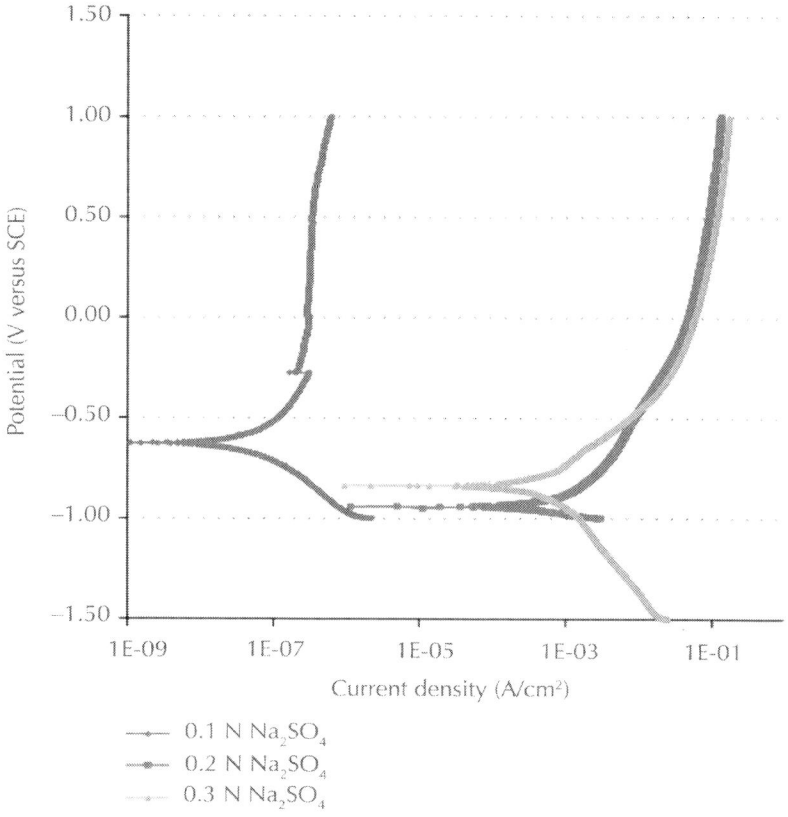

Figure 1: Potentiodynamic polarization curves of low carbon steel in different concentration of Na_2SO_4 solution at pH 6 and temperature 25°C.

Effect of $SO_4^{2-} + S^{2-}$

Addition of S^{2-} to the solution containing SO_4^{2-} further enhances the corrosion rate as can be seen from Figure 2. And the pH has a strong effect on it. It can be seen from (11) and (12) above that the conjoint effect of SO_4^{2-} and S^- is the production of increasing amount of S^-, as well as corrosive sulphur compound.

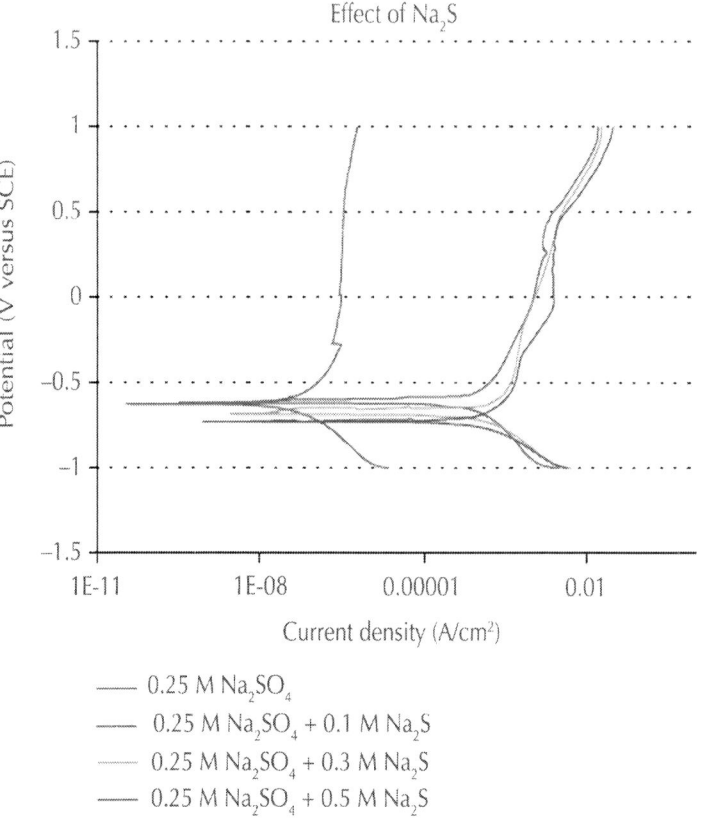

Figure 2: Effect of S^{2-} on potentiodynamic polarization curves of carbon steel in 0.25 M Na_2SO_4 solution at pH 6 and temperature 25°C.

Effect of SO_4^{2-} + Cl^-

Corrosion rate also increases with addition of Cl^- to the solution containing SO_4^{2-} (Figure 3). The rate increases with increase in Cl^- concentration. It is seen from (11)–(14) above that the Cl^- ions attack Fe/Fe^{2+} with the formation of corrosion products $FeCl_2$ and $FeCl_3$. $FeCl_2$ is unstable and may hydrolyse or further react with Cl^- ions to form Fe$(OH)_2$ or $FeCl_3$, respectively.

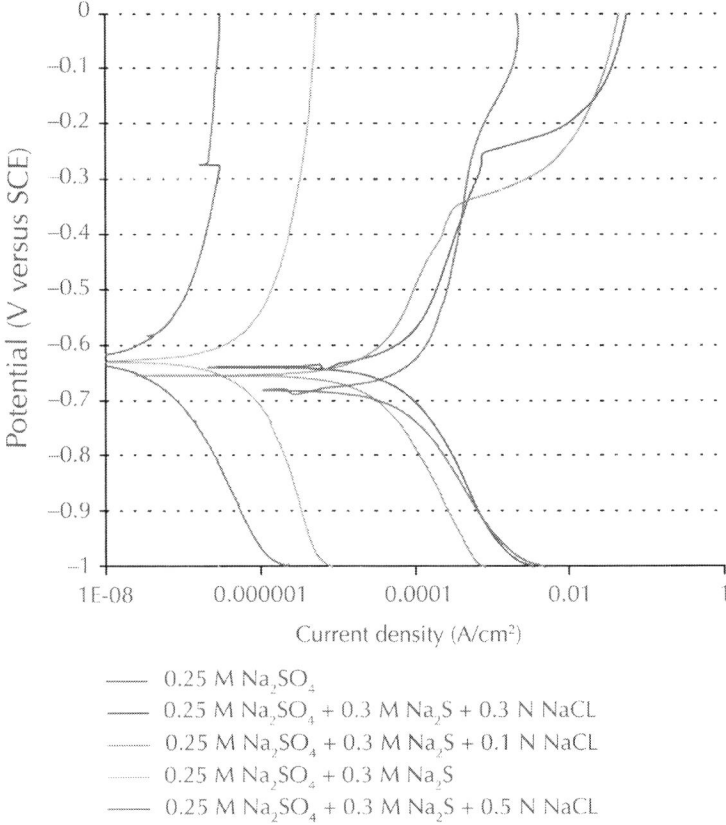

Figure 3: Effect of Cl⁻ on potentiodynamic polarization curves of carbon steel in 0.25 M Na_2SO_4 solution at pH 8 and temperature 25°C.

Effect of SO_4^{2-} + S^{2-} + Cl^-

The conjugate of all 3 ions which are normally present in the oil field sea water is the degradation of carbon steel structure at the highest level. It is seen from Figure 4 that i_{corr} values have shifted to the right and towards the active potential with increasing the concentration of both S^{2-} and Cl^-.

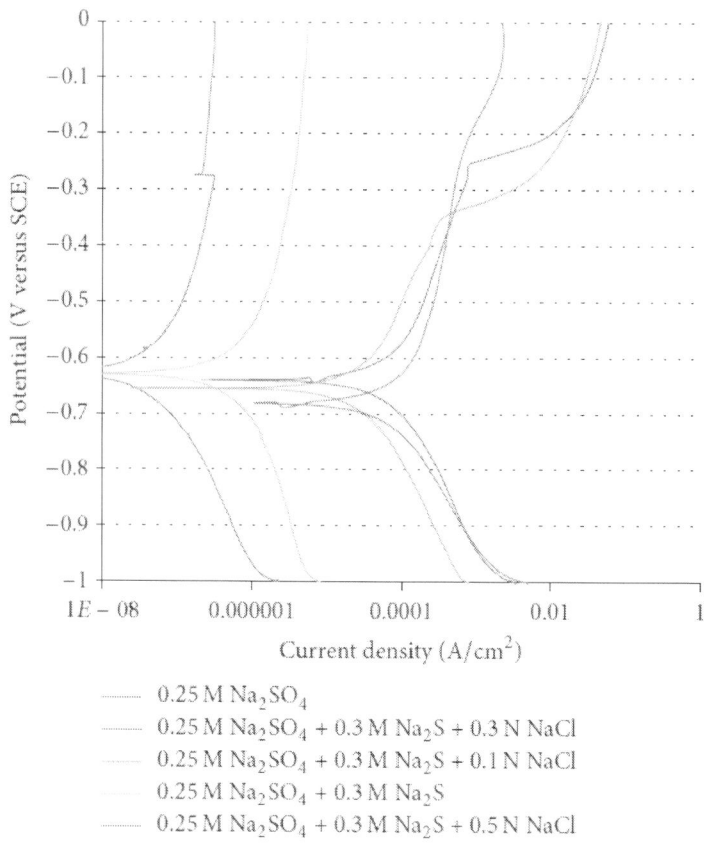

Figure 4: Effect of Cl^- + S^{2-} on potentiodynamic polarization curves of carbon steel in 0.25 M Na_2SO_4 solution at pH 8 and temperature 25°C.

Effect of pH

There is not any much significant effect of polarization curves with change in pH except at pH 11 under alkaline condition (Figure 5), when the corrosion rate is very low. The steel is in the passive region at this pH. The pH of the oil field water is in the range of 4–6 when the corrosion rate is high.

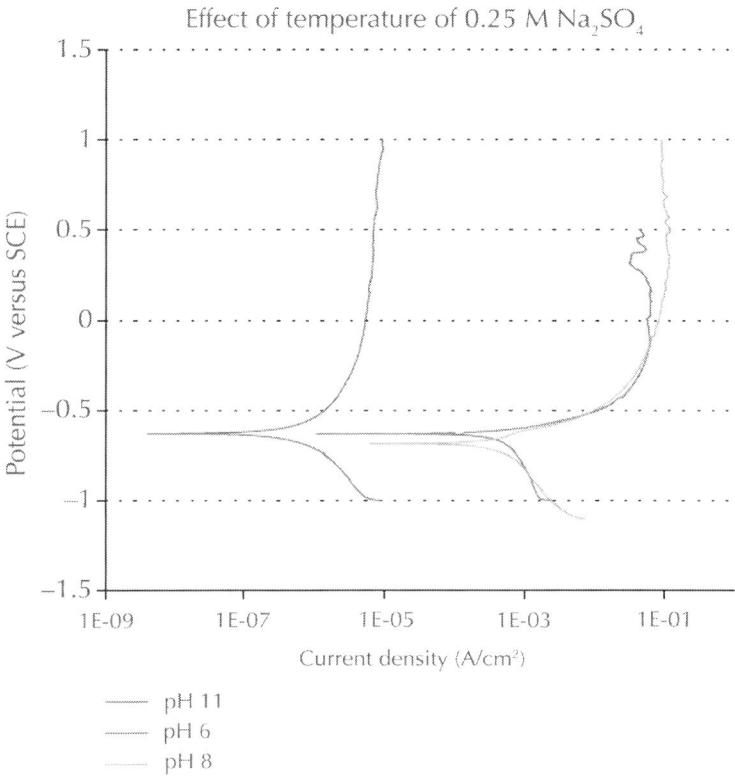

Figure 5: Potentiodynamic polarization curves of carbon steel in an aqueous solution of 0.25 M Na_2SO_4 and 0.3 M Na_2S at temperature 25°C, at different pH.

Effect of Temperature

Temperature aggravates the material degradation (Figure 6) by increasing diffusion and mass transfer coefficient of the aggressive ions corroding the metallic surface. Temperature also increases the I_L, the limiting current density of the concentration polarization, and hence shifts the polarization curves to the right.

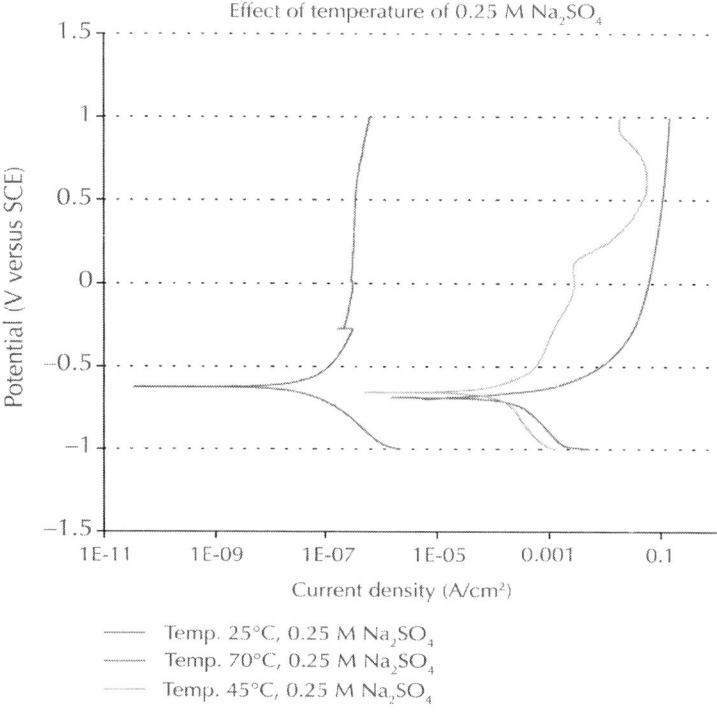

Figure 6: Potentiodynamic polarization curves of low carbon steel in 0.25 M Na_2SO_4 solution at (pH = 8) at different temperature.

Electrochemical Impedance Spectroscopy (EIS)

The phenomenon at the interface of the solid metal and aqueous electrolyte is a complex process consisting of a line of positively and negatively charged ions, capacitance due to double layer, corroded product or film formation on surfaces, polarization resistance (R_p), pore resistance (R_{po}), and various types of impedance due to diffusion of ions, movement of charge in or away from metal surface, and adsorption of cation and anion. The whole phenomenon can be represented by an equivalent AC electrical circuit. The phenomenon can be interpreted from the Bode plots, which are depicted and discussed in the following section for various corrosive species.

Figure 7 displays the Bode plots of carbon steel in solutions of SO_4^{2-}, $SO_4^{2-} + S^{2-}$, and $SO_4^{2-} + S^{2-} + Cl^-$. It is to be noticed that, in all the solutions, EIS data match the model of Randle circuit with a Warburg resistance, W_d (given in the inset of each figure), that prevails at the metal-solution interface. It is seen that the impedance decreases with addition of different aggressive ions compared to those with base solution of only SO_4^{2-}. This decrease in impedance leads to more current flow across the interface and hence increases in corrosion rate. There is a phase angle shift with frequency. The minimum phase angle reaches much less than 90 degrees, indicating the capacitance in the circuit is not a pure capacitance but constant phase element, CPE(Y), and is given by the capacitive impedance equation $Z=1/C (jw)^{-\alpha}$, where α is fraction varying from 0 to 1; the value less than one indicates that it does not behave ideally as pure capacitance. At high frequency, the value of impedance Z_{mod} is roughly equal to R_s, while at low frequency the value is (R_s+R_p). Both can be determined from the blots. Table 1 depicted the computed values of the EIS parameters. It is seen that polarization resistance decreases with addition of more types of ions which support the polarization results of corrosion rates increase as found in Figures 2, 3, and 4. The increase in corrosion rate is supported by the EIS data, increase in Y_o which behaves like capacitance, and decrease in polarization resistance (Table 2). The values of α indicate that the capacitance behaves like constant phase element rather than pure capacitance. The Warburg resistance W_d which signifies resistance to the flow of ions from the solution to corroded metal surface also decreases with addition of more types of ions. All the facts confirm the enhanced corrosion rates of carbon steel in the oil field sea water with presence of $SO_4^{2-} + S^{2-} + Cl^-$.

Table 2: Computed EIS parameters

Sample	Corrosive ions	R_s (ohm)	R_p (ohm)	$Y_o \mu F$	α	W_d (ohm)
Carbon steel	SO_4^{2-}	9.443	8500.0	75.0	0.720	10.430
Carbon steel	$SO_4^{2-} + S^{2-}$	8.268	3700	572.2	0.769	4.187
Carbon steel	$SO_4^{2-} + S^{2-} + Cl^-$	13.31	95	905.6	0.602	2.432

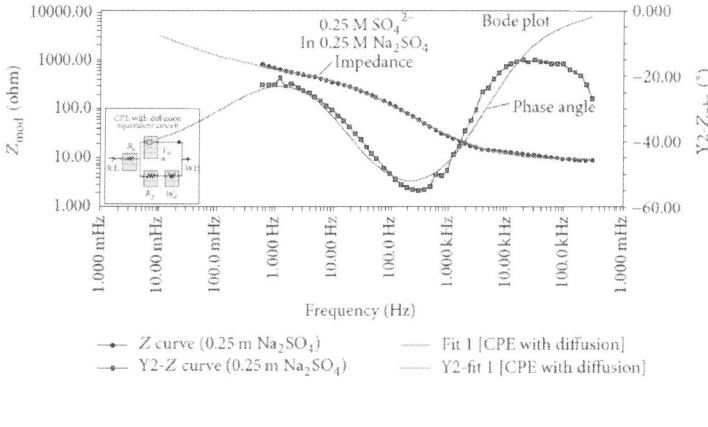

(a)

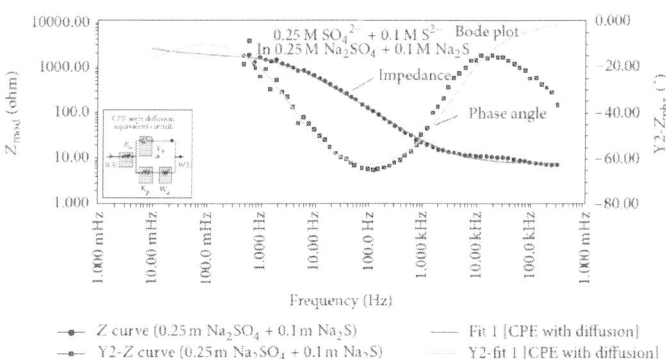

(b)

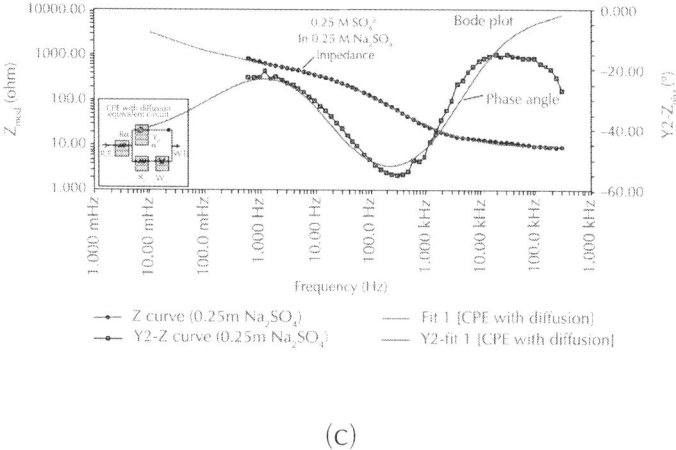

(c)

Figure 7: Bode plots of carbon steel in SO_4^{2-}, $SO_4^{2-} + S^{2-}$, and $SO_4^{2-} + S^{2-} + Cl^-$ solution at pH 8 and temperature 25°C.

X-Ray Diffraction

The presence of corrosion products of FeS, $FeCl_2$, and $FeCl_3$ is clearly indicated by the XRD peak intensities in Figure 8. This supports the corrosion enhancement by SO_4^{2-}, S^{2-}, and Cl^- ions as found in polarization and EIS studies.

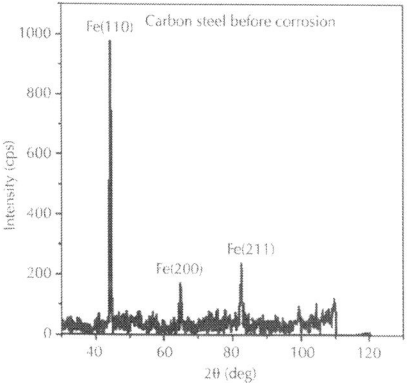

(a)

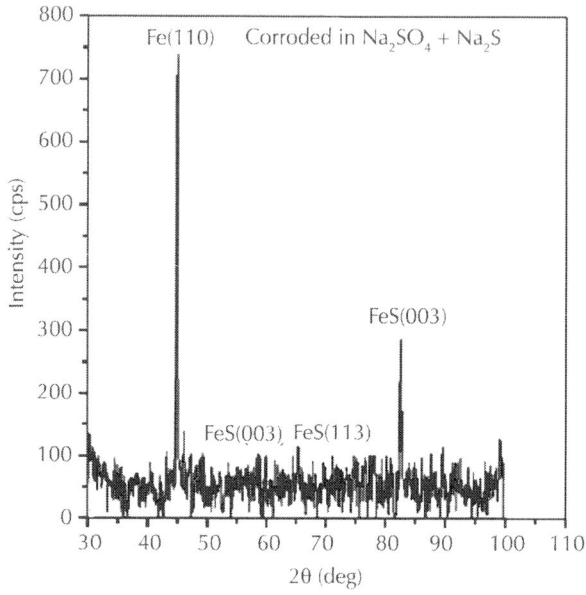

(b)

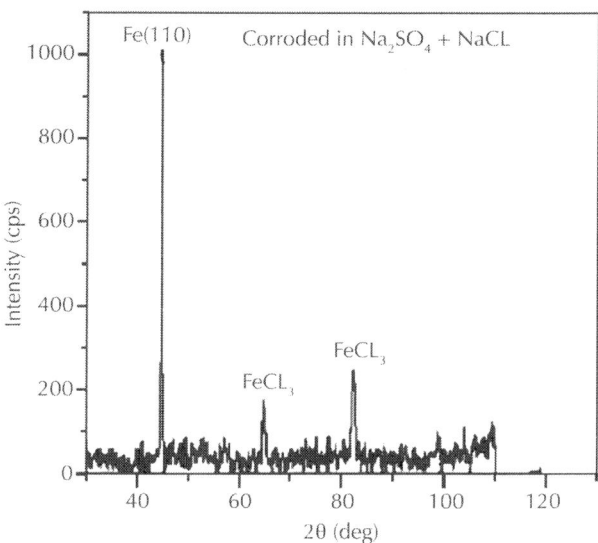

(c)

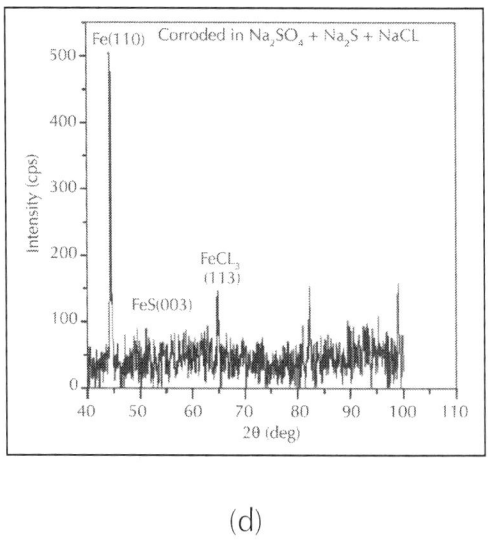

(d)

Figure 8: X-ray diffraction showing peaks of different corroded phases.

Optical and SEM Microscopy Images

The microstructures by optical microscopy clearly show (Figure 9) the corroded structure of steel with products of corrosion over it. The form of corrosion seems to be uniform, not localized. SEM images (Figure10) distinctly reveal how the degree of degradation increases with presence of ions in the solutions from SO_4^{2-} to $SO_4^{2-} + S^{2-}$ to $SO_4^{2-} + S^{2-} + Cl^-$. It is interesting to observe from the morphology of SEM image at higher magnification (Figure 11) that the corrosion has taken place at grain boundary as well as grain body but the attack at grain boundary is very prominent with sulphide (S) corrosion crack. The corrosion products of sulphide compounds as well as element sulphur are clearly revealed. The structure of SEM images shows almost catastrophic failure with the presence of minor and major cracks when the seawater is enriched with the presence of all three corrosive ions, SO_4^{2-}, S^{2-}, and Cl^- The morphology of SEM image analysis along with XRD is in complete agreement with the corrosion data of the polarization experiments and EIS.

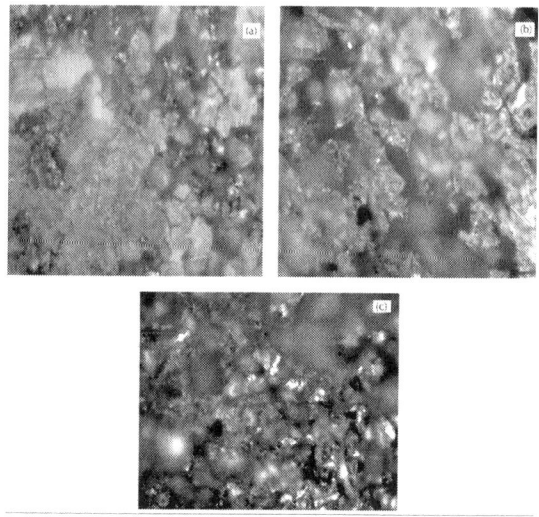

Figure 9: Optical microscopy images of corroded steel (a) Na_2SO_4, (b) $Na_2SO_4 + Na_2S$, and (c) $Na_2SO_4 + Na_2S + NaCl$.

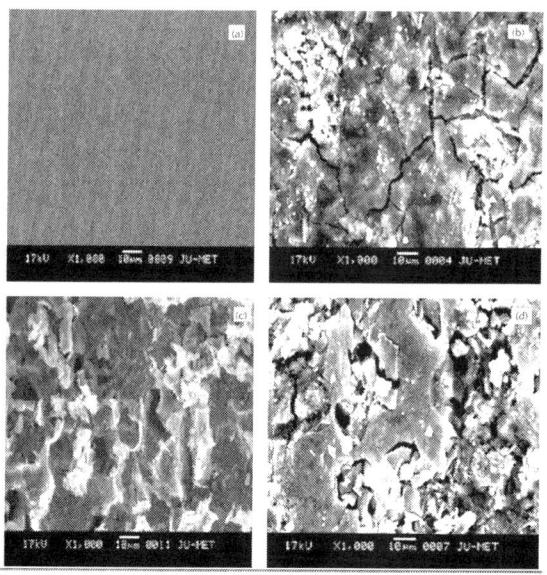

Figure 10: SEM images of corroded steel (a) before corrosion, (b) in Na_2SO_4, (c) in $Na_2SO_4 + Na_2S$, and (d) in $Na_2SO_4 + Na_2S + NaCl$.

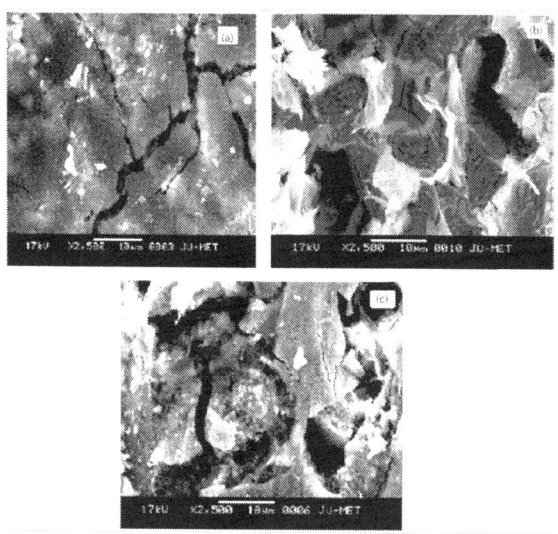

Figure 11: SEM images of corroded steel (a) in Na_2SO_4, (b) in Na_2SO_4 + Na_2S, and (c) in Na_2SO_4 + Na_2S + NaCl.

From (5)–(8), it is seen that SO_4^{2-} is cathodically reduced to give rise to S^{2-}. It is observed from the polarization curve (Figure 1) that the increase of SO_4^{2-} concentration depolarizes the cathodic curves much more compared to anodic ones, shifting E_{corr} towards the negative potential. The release of S^{2-} from the reaction produces corrosion products: iron sulfides (FeS_x) on carbon steel surface. They are nonstoichiometric iron sulfide films mainly composed of mackinawite and pyrrhotite Berner [21–26]. The black color of mackinawite phase is also seen as corrosion products in optical microstructures (Figures 9(a) and 9(b)).

The deterioration of metal due to contact with S^{2-} (H_2S) and moisture is sour corrosion. The forms of sour corrosion are uniform (Figure 9(b)), pitting, and stepwise cracking (Figures 10(c) and 11(b)).

The general equation of sour corrosion can be expressed as follows [21]:

$$S^{2-} + 2H^+ = H_2S\ (aq) \quad E = 0.097\ \text{V versus SCE} \quad (15)$$

Hydrogen sulphide dissociates to produce proton and bisulphide as described [22] by the following equations:

$$H_2S\,(aq) = H^+ + HS^- \quad K = 9.1 \times 10^{-8}$$

$$HS^- = H^+ + S^{2-} \quad K = 9.1 \times 10^{-12} \tag{16}$$

This reaction scheme shows that the presence of hydrogen sulfide can contribute to the concentration of sulfide at the surface by dissociation rather than charge exchange with the surface. The sulfide concentration at the surface is, therefore, dependent on the concentration of aqueous hydrogen sulfide in the electrolyte as well as the reduction processes of sulfate.

H_2S exhibits the different role in anodic process of carbon steel depending on the pH value in the solutions. The local supersaturation of FeS_x could be formed on the carbon steel surface via the following reaction, with the nucleation and growth of one or more of the iron sulfide, mackinawite:

$$H_2S + Fe + H_2O \longrightarrow FeS_x + 2H + H_2O \tag{17}$$

The black corrosion products (FeS_x) formed on the steel surface in the H_2S-containing solutions could be observed (Figure 9(b)). The addition of chloride to the corrosion system did not exhibit any localized corrosion; rather it drastically increased the i_{corr} values and shifted E_{corr} towards the negative potential (Figures 3 and 4), depolarizing the anodic reactions. The increase in i_{corr} with Cl addition seems to have modified the anodic substrate area (Figures 9(c) and 10(d)) and has enhanced the exchange current density for the iron oxidation process.

CONCLUSIONS

All three corrosive ions SO_4^{2-}, S^{2-}, and Cl^- have a strong effect on increasing the corrosion rate of carbon steel.

Cathodic reduction of SO_4^{2-} generates elemental S or S^{2-} ions, in addition to S^{2-} from H_2S in oil. The species cause major corrosion with formation of FeS_x as corrosion products. The attack is found at grain boundary with sulphide cracking as well as some grain body degradation. The effect of addition of Cl^- is the increase of i_{corr} values by hundreds of times possibly due to enhancement of exchange current density of the anodic reactions.

ACKNOWLEDGMENTS

The authors would like to acknowledge "COE, TEQUIP" in Jadavpur University for the support of this work.

REFERENCES

1. S. Nešić, "Key issues related to modelling of internal corrosion of oil and gas pipelines—a review,"Corrosion Science, vol. 49, no. 12, pp. 4308–4338, 2007.
2. L. T. Popoola, A. S. Grema, G. K. Latinwo, B. Gutti, and A. S. Balogun, "Corrosion problems during oil and gas production and its mitigation," International Journal of Industrial Chemistry, vol. 4, aticle 35, 2013.
3. B. W. A. Sherar, P. G. Keech, and D. W. Shoesmith, "The effect of sulfide on the aerobic corrosion of carbon steel in near-neutral pH saline solutions," Corrosion Science, vol. 66, pp. 256–262, 2013.
4. B. W. A. Sherar, I. M. Power, P. G. Keech, S. Mitlin, G. Southam, and D. W. Shoesmith, "Characterizing the effect of carbon steel exposure in sulfide containing solutions to microbially induced corrosion,"Corrosion Science, vol. 53, no. 3, pp. 955–960, 2011.
5. F. M. Song, "A comprehensive model for predicting CO_2 corrosion rate in oil and gas production and transportation

systems," Electrochimica Acta, vol. 55, no. 3, pp. 689–700, 2010.

6. M. Hairil Mohd and J. K. Paik, "Investigation of the corrosion progress characteristics of offshore subsea oil well tubes," Corrosion Science, vol. 67, pp. 130–141, 2013.

7. J. Tang, Y. Shao, T. Zhang, G. Meng, and F. Wang, "Corrosion behaviour of carbon steel in different concentrations of HCl solutions containing H_2S at 90 °C," Corrosion Science, vol. 53, no. 5, pp. 1715–1723, 2011.

8. Z. A. D. Foroulis, "Role of solution pH on wet H_2S cracking in hydrocarbon production," Corrosion Prevention and Control, vol. 40, no. 4, pp. 84–89, 1993.

9. M. A. Veloz and I. González, "Electrochemical study of carbon steel corrosion in buffered acetic acid solutions with chlorides and H_2S," Electrochimica Acta, vol. 48, no. 2, pp. 135–144, 2002.

10. S. Arzola and J. Genescá, "The effect of H_2S concentration on the corrosion behavior of API 5L X-70 steel," Journal of Solid State Electrochemistry, vol. 9, no. 4, pp. 197–200, 2005.

11. H. Y. Ma, X. L. Cheng, S. H. Chen et al., "Theoretical interpretation on impedance spectra for anodic iron dissolution in acidic solutions containing hydrogen sulfide," Corrosion, vol. 54, no. 8, pp. 634–640, 1998.

12. H. Ma, X. Cheng, S. Chen, C. Wang, J. Zhang, and H. Yang, "An ac impedance study of the anodic dissolution of iron in sulfuric acid solutions containing hydrogen sulfide," Journal of Electroanalytical Chemistry, vol. 451, no. 1-2, pp. 11–17, 1998.

13. H. Ma, X. Cheng, G. Li et al., "The influence of hydrogen sulfide on corrosion of iron under different conditions," Corrosion Science, vol. 42, no. 10, pp. 1669–1683, 2000.

14. H.-H. Huang, W.-T. Tsai, and J.-T. Lee, "Electrochemical behavior of the simulated heat-affected zone of A516 carbon steel in H_2S solution," Electrochimica Acta, vol. 41, no. 7-8, pp. 1191–1199, 1996.

15. J. Tang, Y. Shao, J. Guo, T. Zhang, G. Meng, and F. Wang, "The effect of H_2S concentration on the corrosion behavior of carbon steel at 90 °C," Corrosion Science, vol. 52, no. 6, pp. 2050–2058, 2010.
16. P. Smith, S. Roy, D. Swailes, S. Maxwell, D. Page, and J. Lawson, "A model for the corrosion of steel subjected to synthetic produced water containing sulfate, chloride and hydrogen sulfide," Chemical Engineering Science, vol. 66, no. 23, pp. 5775–5790, 2011.
17. R. E. Melchers, "Development of new applied models for steel corrosion in marine applications including shipping," Ships and Offshore Structures, vol. 3, no. 2, pp. 135–144, 2008.
18. M. H. Mohd, D. K. Kim, D. W. Kim, and J. K. Paik, "A time-variant corrosion wastage model for subsea gas pipelines," Ships and Offshore Structures, vol. 9, no. 2, pp. 161–176, 2014.
19. R. E. Melchers and J. K. Paik, "Effect of flexure on rusting of ship›s steel plating," Ships and Offshore Structures, vol. 5, no. 1, pp. 25–31, 2010.
20. M. Shehadeh and I. Hassan, "Study of sacrificial cathodic protection on marine structures in sea and fresh water in relation to flow conditions," Ships and Offshore Structures, vol. 8, no. 1, pp. 102–110, 2013.
21. R. A. Berner, "Thermodynamic stability of sedimentary iron sulfides," The American Journal of Science, vol. 265, pp. 773–785, 1967.
22. R. H. Hausler, L. A. Goeller, R. P. Zimmerman, and R. H. Rosenwald, "Contribution to the "filming amine" theory: an interpretation of experimental results," Corrosion, vol. 28, no. 1, pp. 7–16, 1972.
23. P. Taylor, "The stereochemistry of iron sulfides—a structural ration for the crysta llization of some metastable phases from aqueous solution," American Mineralogist, vol. 65, pp. 1026–1030, 1980.

24. R. A. Berner, "Iron sulfides formed from aqueous solution at low temperatures and atmospheric pressure," The Journal of Geology, vol. 72, pp. 293–306, 1964.
25. D. T. Rickard, "The chemistry of iron sulphide formation at low temperatures," Stockholm Contributions in Geology, vol. 20, pp. 67–95, 1969.
26. A. J. Bard, R. Parsons, and J. Jordan, Standard Potentials in Aqueous Solution, CRC Press, 1st edition, 1985.

Chapter 5

Hydrodynamics of an Inclined Gas–Liquid Cocurrent Up flow Packed Bed

Hana Bouteldja, Mohsen Hamidipour, and
Faïçal Larachi

Department of Chemical Engineering, LAVAL University, Québec, Canada G1V 0A6

ABSTRACT

The effects of inclination on the hydrodynamic behavior of a packed bed operating under gas–liquid cocurrent upflow were experimentally investigated in terms of liquid saturation, bed overall pressure drop and gas–liquid segregation. The non-invasive electrical capacitance tomography (ECT) imaging technique was applied to scrutinize local and axial phase distribution pattern and cross-sectionally averaged liquid saturation. The results indicate that bed inclination creates short circuits for the gas phase along the upper wall where it can flow in a segregated manner. Inception of transition from bubble to segregated flow regime was identified through monitoring a defined uniformity factor for ECT images. Phase segregation developed along the bed with minimum impact in the region close to the entrance. The removed bubbles were replaced by liquid phase resulting in higher liquid saturation values as complete segregation state was approached. The effect of operating conditions on axial profile of liquid saturation was examined.

INTRODUCTION

Bubble columns are used in petrochemical, chemical, pharmaceutical, biochemical and metallurgical industries as multiphase contactors and reactors (Degaleesan et al., 2001). Processes which require good contact between gas and liquid phases can be performed in bubble columns (Prakash and Briens, 1990) where a discontinuous gas phase, in the form of bubbles, circulates upward while accompanying the continuous liquid phase. Bubble columns may also consist of three-phase systems and contain inert, reactive or catalytic particles either in suspension or in a packed bed form. Wastewater treatment, hydrogenation, oxidation, chlorination, polymerization and alkylation are among the processes that have long been performed in bubble column reactors. Other applications such as absorption, catalytic slurry reactions, and coal liquefaction have been also performed in these reactors (Joshi et al., 1990, Blenke, 1979, Chisti, 1989 and Saez et al., 1998). Simplicity of operation, low operating costs, large interfacial area, good inter-phase heat and mass transfers, and ease of adjustment of liquid residence time are the main advantages of this configuration (Kantarcia et al., 2005). However, reliable design and scale-up are known to be restricted by the complex hydrodynamics and its influence on transport characteristics. During the past decades, scientific interest in bubble column reactors has increased considerably (Kantarcia et al., 2005). Research on bubble column covers a wide range of subjects such as gas holdup, bubble properties, flow regimes, back mixing, interfacial area, pressure drop, and heat and mass transfer (Ruzicka et al., 2001, Luther et al., 2004, Majumder et al., 2006, Majumder, 2008 and Dudley, 1995).

Currently, areas of fossil-fuel off-shore extraction and processing are vividly interested on problems linked with inclined multiphase flows. As a matter of fact, the hydrodynamic behavior of floating reactors and separators on embarked boats such as in FLNG (floating liquefied natural gas) and FPSO (floating production, storing and off-loading) systems is a crucial aspect of their design for the prediction of the floating unit performances (Gu and Ju, 2008 and Zhao et al., 2011). Obviously, most hydrodynamic researches in multi-phase reactors have dealt with vertical columns. In contrast, inclined configurations with fixed or slurry catalyst phase are rather on the fringes of studies and are reported very sparsely in the literature. From a pragmatic standpoint, column inclination may have detrimental or beneficial effects on the performance of the reactor depending on its projected utilization. There are a few studies available which investigate the effect of inclination angle of the reactor (O'Dea et al., 1990, Sarkar et al., 1991, Del Pozo et al., 1992, Hudson et al., 1996, Yakubov et al., 2007, Valverde et al., 2008, Atta et al., 2010 and Schubert et al., 2010).

Numerous studies have been reported on the issues associated with gas–liquid and steam–water flows in inclined ducts and channels which are in particular encountered in cooling circuits of pressurized water reactors in nuclear power plants. Singh and Griffith (1970) investigated slug flow of air and water at small upward inclination angles and developed simple correlations for pressure drop and holdup. Slug flow in inclined pipes was also examined by Bonnecaze et al. (1971) who reported data for air–water system. Barnea et al. (1985) reported data on flow pattern transitions for upward gas–liquid flow in pipes at inclination angles from 0° to 90°. Mathematical models previously presented for vertical and horizontal configurations were extended to cover the full range of pipe inclinations.

Flow regimes, fluidization heterogeneity along with the corresponding heat and mass transfer characteristics have been studied in inclined fluidized beds. Arai et al. (1973) theoretically investigated heat exchange between particles and gas in a multistage inclined fluidized bed. A fairly good agreement was recognized between the theoretical and experimental results in a three-stage inclined fluidized bed. Furthermore, effectiveness of multiplying stages was confirmed on heat efficiency both theoretically and experimentally. A chart was made up on the relation between the required number of stages and optimum conditions of heat exchange. Masliyah et al. (1989) studied the enhancement of separation of light and heavy particles from suspensions using inclined channels. They observed that at a fixed set of operating conditions, the increase of inclination from vertical position results in a greater degree of separation. O›Dea et al. (1990) studied inclined fluidized beds, between 45° and 90°, using four different types of powders and air as the fluidizing medium. Flow regimes and transition condition have been identified experimentally and verified by a theoretical model. Sarkar et al. (1991) studied the flow of solid particles from a vertical fluidized bed to a receiving vessel through an inclined downward pipe. They investigated the effects of the connecting pipe length, diameter, and inclination angle and fluidizing agent velocity on the flow rate of solid particles from the fluidized bed to the receiving vessel. Hudson et al. (1996) studied the effects of small inclination angles (up to 10° from the vertical position) in liquid–solid fluidized beds. They measured local holdup and circulation pattern of the liquid and solid phases, and developed a simple model that predicted the solid circulation pattern. Yakubov et al. (2007) studied the structure of a fluidized bed in inclined columns. Experiments were conducted in two glass columns which could be positioned in the whole range of inclination angles, from horizontal to vertical. The results showed that the expansion process of the fluidized bed depends strongly on the inclination angle. The column length was found to have a minor effect on the involved phenomena. Valverde et al. (2008) experimentally examined how bed inclination affects fluidization and sedimentation behavior of fine cohesive particles. They found that the main effect of inclination is to induce fluidization heterogeneity. The local gas velocity increases in the adjacent region to the upper wall at the expense of diminished velocity in the region adjacent to the lower wall. This situation caused early onset of local bubbling in the region adjacent to the upper wall.

The packed bed operated with descending gas–liquid cocurrent flows in slanted configuration was investigated by Schubert et al. (2010) as an extension of studies of horizontal concurrent gas–liquid flows in porous media (Iliuta et al., 2003). In addition to liquid saturation and pressure drop, they studied the transition from segregated/trickle regime to pulse flow regime with respect to inclination angle. The Grosser et al. (1988) flow regime transition model was modified by considering only axial components of the gas and liquid superficial velocities to predict the slant-dependent shifts in transition from segregated/trickle to pulse flow and was found to agree with experimental data. An Eulerian computational fluid dynamics (CFD) framework was implemented by Atta et al. (2010) to simulate an inclined cocurrent downflow packed bed. Two configurations were used (a) a straight tube with an artificially inclined gravity and (b) an inclined geometry with straight gravity. The comparison between electrical capacitance tomography data and the predictions of the liquid saturation showed that there is a considerably quantitative deviation from the experimental data. However, the trends could be satisfactorily predicted. They recommended formulating appropriate drag force closures which should be incorporated in the CFD model for quantifiable approximation of the inclined packed bed hydrodynamics.

Any experimental study analyzing the effect of inclination in packed beds for ascending two-phase flows has been so far fully disregarded. It can be anticipated that inclination would considerably affect the flow patterns and consequently segregated flow may appear in the bed. Thus, knowledge of the basic hydrodynamics for such inclined packed-bed configuration is essential. This study presents the results of an experimental investigation of the hydrodynamic behavior of gas–liquid cocurrent upflow inclined packed bed columns. The goal is to explore the effect of column inclination angle on the flow characteristics. The hydrodynamic response to different operating conditions is studied.

EXPERIMENTAL SETUP

The experimental set-up built to study the effect of inclination angle on two-phase flow hydrodynamics in cocurrent upflow packed bed is illustrated in Fig. 1. Experiments were performed at atmospheric pressure and room temperature. The main element of the set-up is a Plexiglas cylindrical column with an inner diameter (ID) of 0.057 m and a height of 1.50 m. Glass beads (3 mm, =0.395) were used as packing. The packed bed was kept in place using a metallic grid inserted from top to avoid bed fluidization during the experiments. Kerosene and air were used as the liquid and gas phases, respectively. A peristaltic pump was used to feed the liquid from the reservoir to the bottom of the column. The two phases were separated at the top of the column by means of a gas/liquid separator. The liquid was returned to the reservoir and the gas was vented to the atmosphere.

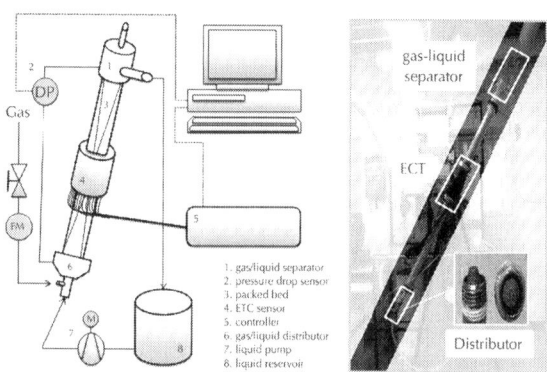

Figure 1: Schematic diagram of the experimental setup.

The column was supported on a steel frame capable of varying the angle of inclination continuously in the full range from horizontal to vertical position. A wide range of gas and liquid superficial velocities were explored to cover the bubble and segregated/bubble flow regimes. Table 1 summarizes the range of operating conditions. The inclination angle was varied in small steps in the range of 0–55°. The angular precision was 0.5°. Pressure drops through the bed was measured using a differential pressure transmitter.

Table 1: System proprieties and range of operating conditions

Parameter	Value/range
Gas superficial velocity, u_G	1.55–46.5 mm/s
Liquid superficial velocity, u_L	0.7–2.8 mm/s
Liquid phase density	792 kg/m³
Liquid viscosity	0.002 kg/m/s
Liquid surface tension	0.025 N/m
Gas density	1.2 kg/m³
Glass beads diameter	3 mm
Sphericity factor	1.0
Bed porosity	0.395
Bed length	1.5 m
Reactor diameter, ID	57 mm
Inclination angle	0–55°

A twin-plane 12-electrode per-plane sensor with a Model DAM200E data acquisition system (PTL300E type) was utilized (Process Tomography, Ltd.) to perform electrical capacitance tomography (ECT) measurements. Tomographic imaging is helpful to capture, without interference with the actual hydrodynamics, the evolving tomography at depths inside the bed that are otherwise inaccessible from mere wall scrutiny. ECT is compatible with oil-like non-polar liquids such as kerosene, as used in our experiments. The ECT system produces instantaneous information with sampling frequencies of 50 Hz. A Tikhonov reconstruction algorithm was chosen to generate normalized permittivity images from the measured capacitances. Selection of this algorithm, with regard to our application was discussed in our previous works (Tibirna et al., 2006 and Hamidipour et al., 2007). The sensor, with an inner diameter of 0.0635 m, can be pushed over the outer wall of the column and fixed at different axial positions. The sensor has been placed at three different axial positions to identify the vertical flow dynamics profiles. ECT calibration enabled setting lowest and highest limits of the normalized permittivities interval so that intermediate per-pixel normalized permittivity values in images during flow conditions can be obtained for rendition of the per-pixel liquid saturations (Hamidipour et al. 2009).

ECT Calibration

Liquid saturation measurements in the packed bed were carried out after the calibration of the ECT sensor at flooded (100%) and drained (0%) bed conditions. The average mixture permittivities can be expressed as

Flooded bed mixture permittivity

e[1]=(1−ε)eS+εeL (1)

Drained-bed mixture permittivity

$$e^{[0]} = (1-\varepsilon)e_S + \left(\varepsilon - \varepsilon_L^{res}\right)e_G + \varepsilon_L^{res} e_L \quad (2)$$

The overall permittivity of a mixture in two-phase operation can be expressed as

Gas–liquid (–solid) mixture permittivity

$$e^{[GL]} = (1-\varepsilon)e_S + \left(\varepsilon - \varepsilon_L^{res} - \varepsilon_L^{fd}\right)e_G + \left(\varepsilon_L^{fd} + \varepsilon_L^{res}\right)e_L \quad (3)$$

where ε, ε_L^{res} and ε_L^{fd} represent, respectively, the bed porosity, the residual liquid holdup retained due to capillary forces and free-draining liquid holdup, and eS, eL, and eG, are the packing, liquid, and gas permittivities, respectively. Furthermore, the free-draining liquid holdup normalized by the effective bed porosity (after resting the residual liquid holdup) or the free-draining liquid saturation (β_L^{fd}), can be estimated using the normalized permittivity, NoP

$$NoP = \frac{e^{[GL]} - e^{[0]}}{e^{[1]} - e^{[0]}} = \frac{\varepsilon_L^{fd}}{\varepsilon - \varepsilon_L^{res}} = \beta_L^{fd} \quad (4)$$

Instantaneous local free-draining saturation values for all the 32×32 pixels of the ECT image were determined, β_L^{fd}, to provide the cross-sectionally averaged free draining liquid saturation, β_L^{fd},

$$\beta_L^{fd} = \frac{1}{NP}\sum_{i=1}^{NP} \beta_{L,i}^{fd} \quad \text{and} \quad \overline{\beta_L^{fd}} = \frac{1}{NF}\sum_{j=1}^{NF}\left(\frac{1}{NP}\sum_{i=1}^{NP}\beta_{L,i}^{fd}\right) \quad (5)$$

Time-averaged liquid saturation values, $\overline{\beta_L^{fd}}$, were obtained for steady-state flow conditions from ~300 successive cross-sectional images. Number of pixels and frames applied are denoted by NP and NF, respectively. Afterward, using free-draining liquid saturations (Eq. (4)), the free-draining liquid holdup, ε_L^{fd}, could be calculated knowing the experimentally determined bed porosity, ε, and the residual liquid holdup, ε_L^{res}.

RESULTS AND DISCUSSION

Effect of Inclination Angle on Local Liquid Distribution Pattern and Overall Bed Pressure Drop

The study was realized through examination of liquid saturation and bed pressure drop as a function of inclination angle and fluid throughputs. The inclination angle was initiated from uniform liquid distribution (i.e., 0°) and changed up to 55° to cover a wide range of inclinations. To characterize the flow patterns that evolve from bubble to segregated flow regime, the non-invasive ECT method was applied. Each ECT image represents a two-dimensional tomogram averaged over 5 cm thick slices in the reactor corresponding to the electrodes› heights. Systematic experimental comparisons between vertically aligned packed bed reactor and inclined bed in two-phase flow operation were carried out to determine at which inclination angle deviations in flow patterns start to occur. Therefore, the conventional straight reactor operated in a stable bubble flow regime was gradually inclined. Fig. 2 shows the development of the liquid saturation over the cross section of the packed bed 55 cm away from bed entrance for inclinations varying from θ=0° (vertical) up to 55°. Vessel inclination inevitably affects the gas–liquid distribution and flow patterns inside the packed bed. At 10°, the segregation has already started to develop becoming more pronounced the higher the inclination. The appearance of blue color in ECT images represents the formation of gas channels along the upper wall consisting of large bubbles as a function of increased inclination angle resulting in less gas–liquid contact. The red color illustrates that liquid is moving toward a gas-free state, i.e., complete segregation, alongside the lower wall.

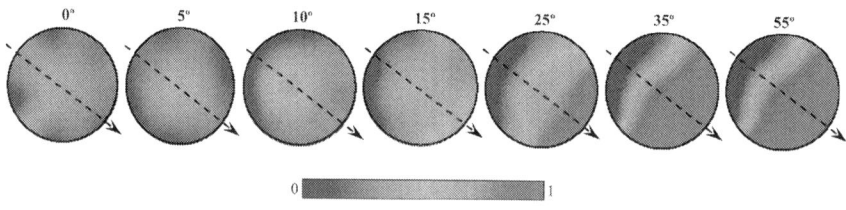

Figure 2: Exemplary illustration of the inclination effect on the flow texture (u_G=15.5 mm/s; u_L=0.7 mm/s, 55 cm away from the bed entrance). (For interpretation of the references to color in this figure legend, the reader is referred to the web version of this article).

The segregation state can be identified based on the lack of crosswise uniformity of the liquid saturation distribution. A uniformity factor, χ, was defined based on the deviation of pixel saturation with respect to the average cross-sectional liquid saturation

$$\chi = \frac{1}{NF} \sum_{j=1}^{NF} \left(\frac{1}{NP} \sum_{i=1}^{NP} \left(\frac{\beta_{L,i}^{fd} - \beta_L^{fd}}{\beta_L^{fd}} \right)^2 \right)$$
(6)

Where NP is the number of pixels in the image, NF is the number of successive cross-sectional images (~300 frames) after establishment of a steady state. Also, $\beta_{L,i}^{fd}$ and β_L^{fd} are the ith pixel and average cross-sectional liquid saturations for each image, respectively. While close to zero value is an indication of a uniform distribution, higher uniformity factor values describe occurrence of mal-distribution. The uniformity factor values for different operating conditions as a function of inclination angle at 55 cm away from the bed entrance are shown in Fig. 3. For the presented range of operating conditions, the effect of inclination angle on the uniformity factor starts at ~10°. All operating conditions reveal a peak of uniformity factor where the worst distribution pattern takes place. As the bed is tilted the gas phase has the tendency to escape toward the upper wall. At a critical angle, complete phase segregation starts to develop and more cross-sectional area is uniformly covered by either liquid or gas phases. Hence, the uniformity factor diminishes. Therefore, the peak characterizes the transition from bubble to segregated flow regime. Fig. 3 depicts that the location of peak depends on the gas and liquid velocities. At low phase interaction (i.e., low superficial velocity) segregation requires smaller inclination angle to establish (~25°). At high phase interaction (i.e., high superficial velocity) more deviation from vertical position is required for segregation to evolve (~35°). Liquid phase appears to play an important role. Fig. 3 shows that higher liquid velocity prevents facile escape of gas bubbles toward the upper wall, therefore, segregation occurs at higher inclination angle.

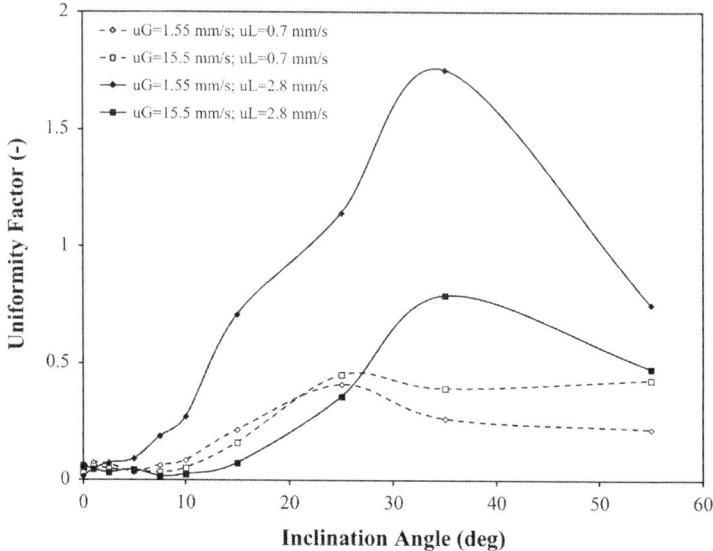

Figure 3: Effect of inclination angle on uniformity factor (55 cm away from the bed entrance).

The cross-sectionally averaged liquid saturation (Eq. 5, at 55 cm away from the bed entrance) data were analyzed as well as the bed overall pressure drop (Fig. 4a, b). The liquid saturation (Fig. 4a) is clearly affected by the inclination angle for all superficial fluid velocities investigated in this study. At vertical position, increasing gas velocity at constant liquid throughput results in higher presence of gas phase which in turn causes lower liquid saturation. For the selected range of operating conditions, at around 10° deviation from vertical position, the average liquid saturation started to increase. As the bed is inclined, gas and liquid start to segregate. The space originally occupied by gas is filled by liquid forcing the gas phase to squeeze close to the upper wall. Higher inclination angles, approaching complete segregation state, cause more coverage of cross-sectional area by the liquid. Therefore, higher liquid saturation is observed. As mentioned earlier, liquid velocity has an important role to maintain bubbles in the liquid phase. Consequently, at same inclination angle, higher liquid velocity results in lower liquid saturation.

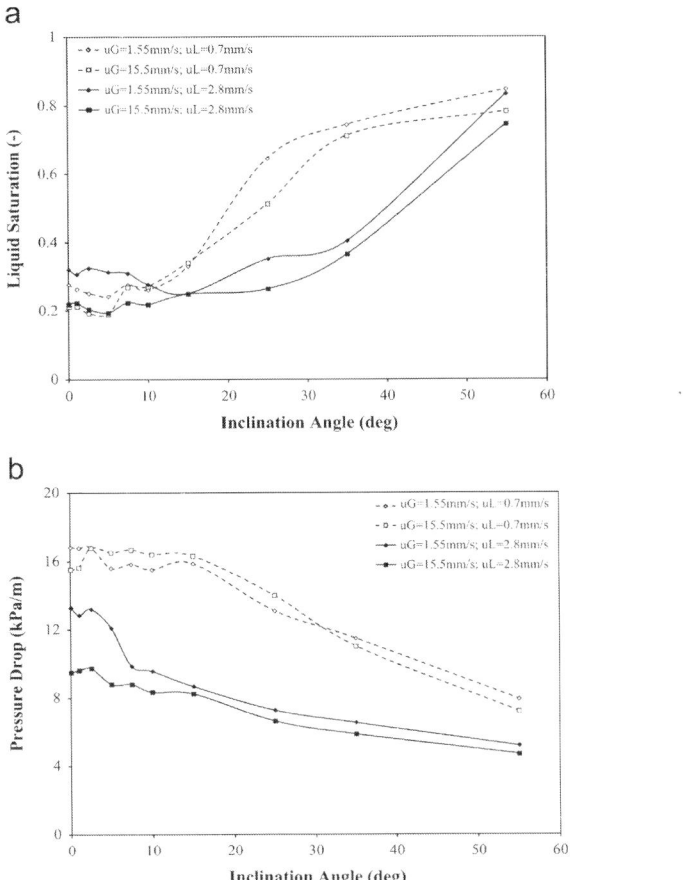

Figure 4: Effect of inclination angle on (a) on cross-sectional average liquid saturation (55 cm away from the bed entrance) and (b) overall bed pressure drop.

In a bubble column, two main factors contribute to the overall bed pressure drop: (i) the interaction between gas, liquid and solid phases (i.e., drag forces) and (ii) the static head of liquid phase. Fig. 4b shows the variation of bed overall pressure drop as a function of inclination angle. Till around 10° inclination, as the distribution pattern does not change significantly, a minor effect is observed on pressure drop. Further inclination creates an easy path for the gas phase close to the upper wall resulting in lower phase interactions. Therefore, as the bed is tilted toward complete segregation, pressure drop continues to decline. At lower liquid superficial velocity gas holdup is lower (Fig. 4a) and the contribution of higher static liquid head results in greater overall pressure drop. The results show that for the selected range of velocities at vertical position, for a constant gas superficial velocity, pressure drop decreases by increase of liquid velocity. Higher liquid throughput causes more phase interaction when the two phases compete for flow path. However, a greater drag force is imposed on the gas phase by a low velocity liquid phase. The net outcome for the range of our experiments (i.e., low velocities to prevent pulse flow regime) is less overall bed pressure drop.

Effect of Inclination Angle and Operating Conditions on the Axial Phase Distribution

An Eulerian slice representation of ECT images is used to visualize the axial development of the liquid-flow field as shown in Fig. 5a, b (Hamidipour and Larachi, 2010). In our case, Eulerian slices were plotted to study the morphology of two-phase flow under segregated/bubble flow regime (Fig. 6a–c). Pixelized liquid saturations reconstructed along a selected diametrical line (e.g., A–A line in Fig. 5a) was shot at 50 Hz pace as recorded and plotted one after the other from individual images recorded during two-phase flow operation. In the current instance, the perpendicular line A–A crossing the segregated liquid area from the bottommost to the uppermost wall areas at a given axial position is the most representative line. Evolving this line time-wise from bottom to top as in Fig. 5a, would depict upstream events for the flow direction but delayed in time until they hit the tomograph sensing plane. This gives, in an approximate sense, virtual local axial tomogram representation of the liquid-flow field.

Hydrodynamics of an Inclined Gas–Liquid Cocurrent Up flow... 163

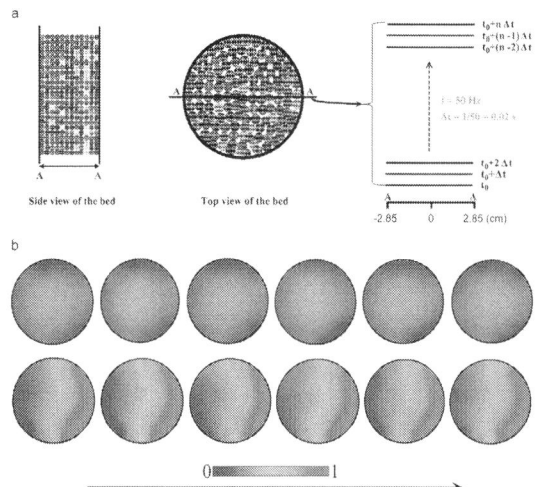

Figure 5: (a) Schematic of Eulerian slice construction, and (b) ECT snapshots of bubble and segregated flow regimes; first row at vertical position, 120 cm away from the bed entrance, u_L=1.4 mm/s; u_G=1.55 mm/s; second row 55 cm away from the bed entrance, inclination angle=25°, u_L=1.4 mm/s; u_G=1.55 mm/s.

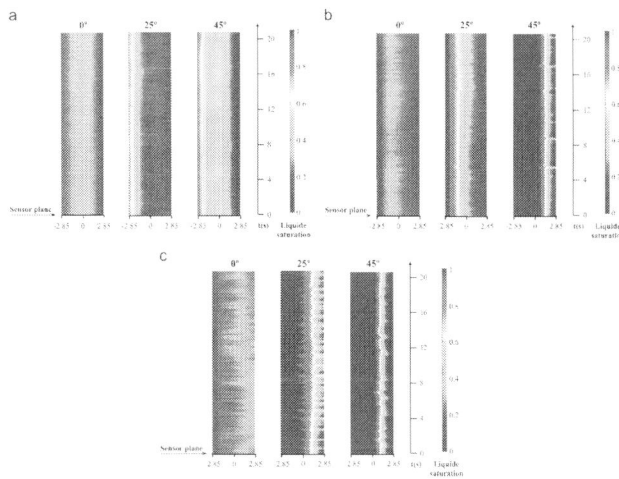

Figure 6: Eulerian slice representation for u_L=1.4 mm/s; u_G=1.55 mm/s, (a) 20 cm, (b) 55 cm, and (c) 120 cm away from the bed entrance.

Fig. 5b shows typical images used to build the Eulerian slices. Each row shows different instants of flow under specific operating conditions. The first row (Fig. 5b) was obtained at vertical position, 120 cm away from the bed entrance (u_L=1.4 mm/s; u_G=1.55 mm/s). The second row presents different moments of flow, 55 cm away from of the bed entrance while bed was inclined at 25° (u_L=1.4 mm/s; u_G=1.55 mm/s). Several images (21 s×50 Hz) similar to these two rows were used to construct Fig. 6a–c (i.e., Eulerian slices). Liquid and gas superficial velocites were kept constant (u_L=1.4 mm/s; u_G=1.55 mm/s) whilst the inclination angle was increased stepwise (0°, 25°, and 45°) for different axial positions of ECT, 20 cm, 55 cm and 120 cm away from bed entrance.

Maintaining the bed vertically revealed no effect of height. Under bubble flow regime and vertically positioned bed (°0) slightly lower liquid saturations are observed close to the walls. This is attributed to the higher porosity in this region. As inclination angle was increased, the gravity force acting on the liquid increased resulting in more accumulation of liquid in the bottom wall region. Lower liquid saturation nearby the upper wall area is in accordance with the passage of gas bubbles. It is noted that at inclination angle equal to 25° middle of the column shows less phase segregation compared to the top of the column. At 45°, segregation is less close to the entrance of the column; however, the distribution pattern from middle to the top of column is similar. In fact, segregation requires a minimum length to develop which is a function of inclination angle. At higher inclination angles less distance is necessary to come close to a complete phase segregation state. At the top of the column, a wavy pattern is observed at an inclination angle equal to 25°. The waves tend to disappears with the increase of inclination angle (Fig. 6c). The waves are an indication of the region where gas–liquid disengagement occurs. At higher inclination angles this region moves toward the bed entrance, therefore, a relativly calm pattern is established downstream of the disengagement zone.

Fig. 7a shows the effect of inclination angle on the axial profile of cross-sectionally averaged liquid saturation for u_G=1.55 mm/s and u_L=2.8 mm/s. At vertical position (i.e., 0°) liquid saturation is independent of bed height. An increasing trend of liquid saturation is observed along the bed with deviations from vertical position. Development of phase segregation along the bed creates more space for liquid presence equivalent to higher liquid saturation. For the selected operating conditions, at inclination angle=45°, final phase distribution pattern was approached around midway of the bed resulting in an almost invariant liquid saturation afterwards. It is important to note that even for small inclination angles complete segregation might take place in a long bed.

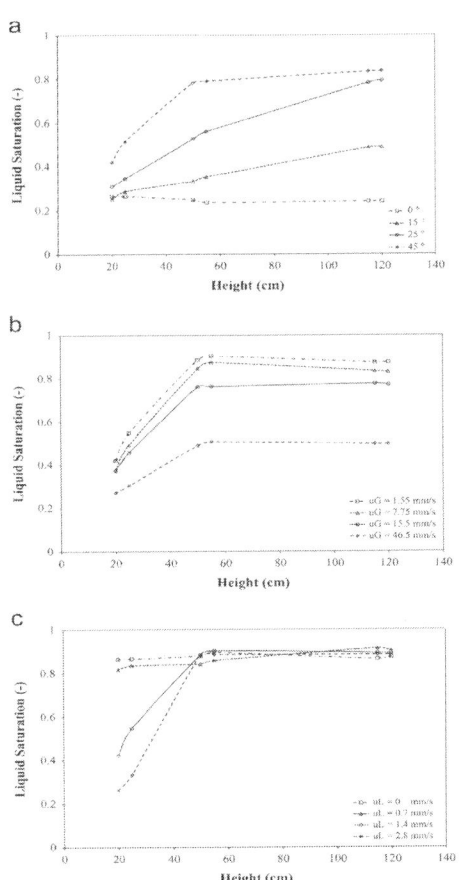

Figure 7: Axial profile of liquid saturation, (a) effect of inclination angle, u_G=1.55 mm/s and u_L=2.8 mm/s, (b) effect of gas superfi-

cial velocity,u_L=2.8 mm/s and inclination angle=45°, and (c) effect of liquid superficial velocity, u_G=1.55 mm/s and inclination angle=45°.

Fig. 7b depicts the effect of gas superficial velocity on the axial profile of liquid saturation for u_L=2.8 mm/s and inclination angle=45°. Similar to the vertical bed behavior, higher gas velocity boosts the gas–liquid interaction while seeking flow space. Thus, part of liquid is pushed out and lower liquid saturation is observed along the bed. This effect is less pronounced in the region close to the entrance due to higher static head of liquid phase which imposes higher resistance.

The influence of liquid superficial velocity on the axial profile of liquid saturation for u_G=1.55 mm/s and inclination angle=45° is presented in Fig. 7c. At low liquid velocity, no significant effect is observed along the bed. Increase of liquid velocity results in lower liquid saturation close to bed entrance. The axial shear imposed on the gas bubbles by the liquid phase retards bubble removal to higher elevations which in turn lowers the area occupied by liquid (i.e., less liquid saturation).

CONCLUSIONS

The hydrodynamic behavior of an inclined gas–liquid cocurrent upflow packed bed was experimentally investigated. The inclination angle was varied from 0° (i.e., vertical position) up to 55° to cover a wide range of inclination. The electrical capacitance tomography (ECT) imaging technique was implemented to monitor local and axial distribution patterns. Cross-sectionally averaged liquid saturation, bed overall pressure drop and gas–liquid segregation were measured. In addition, the effect of operating conditions on axial profile of liquid saturation was examined. The experimental results indicated that

–Short circuits of gas phase were formed along the upper wall due to bed inclination. Thus, phase interaction decreased causing lower pressure drop values.

–Inception of transition from bubble to segregated flow regime was identified through monitoring uniformity factor for ECT images. The worst distribution pattern (i.e., the highest uniformity factor) was considered as the flow regime transition point.

–Segregation was developed along the bed and removed bubbles were replaced by liquid phase resulting in higher liquid saturation values as complete segregation state was approached.

–High liquid superficial velocity imposed a greater shear on the gas bubbles preventing segregation especially in the region close to the entrance.

ACKNOWLEDGMENTS

Financial support from the Canada Research Chair "Green processes for cleaner and sustainable energy" and the Discovery Grant from the Natural Sciences and Engineering Research Council (NSERC) is gratefully acknowledged.

REFERENCES

1. Arai, N., Hasatani, M., Sugiyama, S., 1973. Heat transfer in multistage inclined fluidized beds. Chemical Engineering 37, 379.
2. Atta, A., Schubert, M., Nigam, K.D.P., Roy, S., Larachi, F., 2010. Co-current descending two-phase flows in inclined packed beds: experiments versus simulations. Canadian Journal of Chemical Engineering 88, 742–750.
3. Bonnecaze, R.H., Erskine, W., Greskovich, E.J., 1971. Hold-up and pressure drop for two-phase slug flow in inclined pipelines. AIChE Journal 17, 1109–1113.
4. Blenke, H., 1979. Loop reactors. Advances in Biochemical Engineering 13, 121–215.

5. Barnea, D., Shoham, O., Taitel, Y., 1985. Gas–liquid flow in inclined tubes: flow pattern transitions for upward flow. Chemical Engineering Science 40, 131–136.
6. Chisti, Y., 1989. Airlift Bioreactors. Elsevier Applied Science, London and New York.
7. Del Pozo, M., Briens, C.L., Wild, G., 1992. Effect of column inclination on the performance of three-phase fluidized beds. AIChE Journal 38, 1206–1212.
8. Dudley, J., 1995. Mass transfer in bubble columns: a comparison of correlations. Water Research 29, 1129–1138.
9. Degaleesan, S., Dudukovic, M., Pan, Y., 2001. Experimental study of gas-induced liquid-flow structures in bubble columns. AIChE Journal 47, 1913–1931.
10. Grosser, K., Carbonell, R.G., Sundresan, S., 1988. Onset of pulsing in two-phase cocurrent downflow through a packed bed. AIChE Journal 34, 1850–1860.
11. Gu, Y., Ju, Y., 2008. LNG-FPSO: offshore LNG solution. Frontiers of Energy and Power Engineering in China 2, 249–255.
12. Hamidipour, M., Larachi, F., Ring, Z., 2007. Cyclic operation strategy for extending cycle life of trickle beds under gas liquid filtration. Chemical Engineering Science 62, 7426–7435.
13. Hamidipour, M., Larachi, F., Ring, Z., 2009. Monitoring filtration in trickle beds using electrical capacitance tomography. Industrial and Engineering Chemistry Research 48, 1140–1153.
14. Hamidipour, M., Larachi, F., 2010. Characterizing the liquid dynamics in concurrent gas–liquid flows in porous media using twin-plane electrical capacitance tomography. Chemical Engineering Journal 165, 310–323.
15. Hudson, C., Briens, C.L., Prakash, A., 1996. Effect of inclination on liquid solid fluidized beds. Powder Technology 89, 101–113.

16. Iliuta, I., Fourard, M., Larachi, F., 2003. Hydrodynamic model for horizontal twophase flow through porous media. Canadian Journal of Chemical Engineering 81, 957–962.
17. Joshi, J.B., Ranade, V., Gharat, S.D., Lele, S.S., 1990. Sparged loop reactors. Canadian Journal of Chemical Engineering 68, 705–741.
18. Kantarcia, N., Borakb, F., Ulgena, K.O., 2005. Bubble column reactors. Process Biochemistry 40, 2263–2283.
19. Luther, S., Rensen, J., Guet, S., 2004. Bubble aspect ratio and velocity measurement using a four-point fiber-optical probe. Experiments in Fluids 36, 326–333.
20. Majumder, S.K., 2008. Analysis of dispersion coefficient of bubble motion and velocity characteristic factor in down and upflow bubble column reactor. Chemical Engineering Science 63, 3160–3170.
21. Majumder, S.K., Kundu, G., Mukherjee, D., 2006. Prediction of pressure drop in a modified gas–liquid down flow bubble column. Chemical Engineering Science 61, 4060–4070.
22. Masliyah, J.H., Nasr-El-Din, H., Nandakumar, K., 1989. Continuous separation of bidisperse suspensions in inclined channels. International Journal of Multiphase Flow 15, 815–829.
23. O'Dea, D.P., Rudolph, V., Chong, Y.O., Leung, L.S., 1990. The effect of inclination on fluidized beds. Powder Technology 63, 169–178.
24. Prakash, A., Briens, C.L., 1990. Porous gas distributors in bubble columns: effect of start-up procedure on distributor performance. Canadian Journal of Chemical Engineering 68, 204–210.
25. Ruzicka, M.C., Zahradnık, J., Drahos, J., Thomas, N.H., 2001. Homogeneous heterogeneous regime transition in bubble columns. Chemical Engineering Science 56, 4609–4626.
26. Saez, A.E., Marquez, M.A., Roberts, G.W., Carbonell, R.G., 1998. Hydrodynamic model for gas-lift reactors. AIChE Journal 44, 1413–1423.

27. Sarkar, M., Gupta, S.K., Sarkar, M.K., 1991. An experimental investigation of the flow of solids from a fluidized bed through an inclined pipe. Powder Technology 64, 221–231.
28. Schubert, M., Hamidipour, M., Duchesne, C., Larachi, F., 2010. Hydrodynamics of cocurrent two-phase flows in slanted porous media – modulation of pulse flow via bed obliquity. AIChE Journal 56, 3189–3205.
29. Singh, G., Griffith, P., 1970. Determination of the pressure drop optimum pipe size for a two-phase slug flow in an inclined pipe. Journal of Manufacturing Science and Engineering 92, 717–726.
30. Tibirna, C., Edouard, D., Fortin, A., Larachi, F., 2006. Usability of ECT for quantitative and qualitative characterization of trickle-bed flow dynamics subject to filtration conditions. Chemical Engineering and Processing 45, 538–545.
31. Valverde, J.M., Castellanos, A., Quintanilla, M.A.S., Gilabert, F.A., 2008. Effect of inclination on gas-fluidized beds of fine cohesive powders. Powder Technology 182, 398–405.
32. Yakubov, B., Tanny, J., Maron, D.M., Brauner, N., 2007. The dynamics and structure of a liquid–solid fluidized bed in inclined pipes. Chemical Engineering Journal 128, 105–114.
33. Zhao, W.H., Yang, J.M., Hu, Z.Q., Wei, Y.F., 2011. Recent developments on the hydrodynamics of floating liquid natural gas (FLNG). Ocean Engineering 38, 1555–1567.

Chapter 6

Interaction between Aqueous Solutions of Hydrophobically Associating Polyacrylamide and Dodecyl Dimethyl Betaine

Zhongbin Ye[1,2], Guangfan Guo[3], Hong Chen[1], and Zheng Shu[3]

[1]State Key Laboratory of Oil and Gas Reservoir Geology and Exploitation, Southwest Petroleum University, Chengdu, Sichuan 610500, China
[2]School of Chemistry and Chemical Engineering, Southwest Petroleum University, Chengdu, Sichuan 610500, China
[3]School of Petroleum Engineering, Southwest Petroleum University, Chengdu, Sichuan 610500, China

ABSTRACT

The interaction between hydrophobically associating polyacrylamide (HAPAM) and dodecyl dimethyl betaine (BS-12) is studied through surface tension, interfacial tension (IFT), apparent viscosity, aggregation behavior, and microscopic morphologies. Results show that surface and interface properties of BS-12 are largely affected by HAPAM. BS-12 critical micelle concentrations are increased with the increment of polymer concentrations. Abilities of reduced air-water surface tension and oil-water interfacial tension are dropped. The oil-water interfacial tension to reach minimum time is increased. HAPAM can form network structures in the aqueous solution. Mixed micelles are formed by the interaction between BS-12 micelles and hydrophobic groups of HAPAM in aqueous solution and self-assembly behavior of HAPAM is affected. With the increment of surfactant concentrations, the apparent viscosity, apparent weight average molecular weights (M_w), root mean square radius of gyration ($\langle R_g \rangle$), and hydrodynamic radius of HAPAM increase first and then decline. Moreover, microscopic morphologies of the mixed system are formed from relatively loose network structures to dense network structures and then become looser network structures and the part of network structures fracture.

INTRODUCTION

Water-soluble polymers modified with a small amount of hydrophobic groups (<2%, mole fraction) have become of great interest in recent years. They have broad application prospects, such as oil exploration, paint, mineral separation, and cosmetic and pharmaceutical preparations et al. [1–4]. Hydrophobically associating polymer molecules in aqueous solution, three-dimensional network structures are formed through the intermolecular interaction. Therefore, hydrophobically modified polymers have the good increasing viscosity, heat resistance, salt resistance and shear resistance, and so on [5–7], and they have a good application prospect.

Polymer-surfactant mixed system is a kind of very important soft substance. Usually adding surfactants into polymer solutions, the dosage of polymers or surfactants can be reduced, and solution performances are improved [8–10]. The mixed system has many unique properties, for example, the system viscosity, interfacial adsorption, solubilization, drug delivery, and so on. So the study of polymer-surfactant mixed system is that people are very interested in research subjects [11–14]. In recent years, the interaction between hydrophobically associating polyacrylamide (HAPAM) and surfactant is studied by some scholars [15–18]. These studies find that the system can form mixed micelles and rheological properties of polymer solutions can be changed to a great extent. The interaction between polymer and surfactant is mainly decided by the polymer hydrophobicity and the surfactant structure. Betaine is a main type of zwitterionic surfactant. It shows characteristics of different ionic surfactant under different conditions owing to having anionic and cationic groups at the same time. Now the interaction between HAPAM and betaine surfactants is very little researched [19]. In this paper, the interaction of hydrophobically associating polyacrylamide (HAPAM) and dodecyl dimethyl betaine is studied. Expecting polymer-surfactant mixed system has good application prospects in washing textile, daily chemical and oil field development, and so on.

EXPERIMENT

Materials and Instruments

Acrylamide (AM), acrylic acid (AA), anhydrous sodium carbonate (Na_2CO_3), potassium persulfate ($K_2S_2O_8$), and sodium hydrogen sulfite ($NaHSO_3$) were provided by Chengdu Kelong Chemical Reagents Corporation, China, and all drugs were analytical pure.

18-Alkyl-dimethyl diallyl ammonium chloride (C_{18}DMAAC, its molecular structure be shown in Figure 1) and dodecyl dimethyl betaine (BS-12) were supplied by Southwest Petroleum University. TX-500C full range tensiometer was purchased from the United States Bowing Industry Corporation. Brookfield DV-III viscometer was purchased from the United States Brookfield Instrument Corporation. BI-200SM dynamic/static wide angle laser light scattering apparatus was purchased from the United States Brooke Haven Instrument Corporation. Nanoscope IIIa atomic force microscope was purchased from the United States Digital Instrument Corporation.

$$H_2C=CH-CH_2-\overset{\overset{CH_3}{|}}{\underset{\underset{CH_3}{|}}{N}}\overset{\ominus}{Cl}-(CH_2)_{17}-CH_3$$

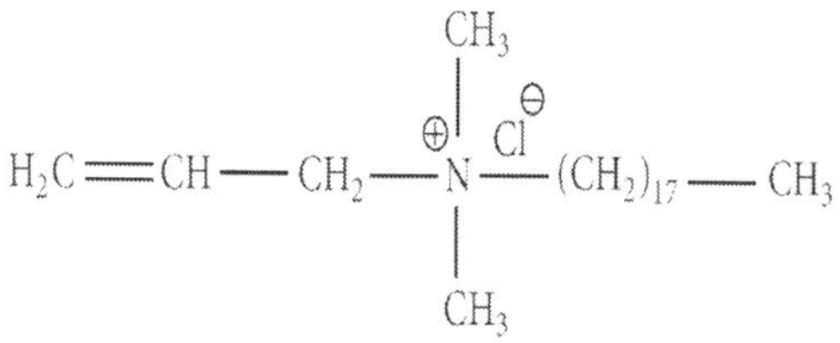

Figure 1: Molecular structure of 18-alkyl-dimethyl diallyl ammonium chloride (C_{18}DMAAC).

Synthesis of Hydrophobically Associating Polyacrylamide

Hydrophobically associating polyacrylamide (HAPAM) synthesis is as follows. 80 g distilled water, 14.9 g acrylamide (AM), 5 g sodium acrylate (acrylic acid and anhydrous sodium carbonate), and 0.1 g 18-alkyl-dimethyl diallyl ammonium chloride (C_{18}DMAAC) were added into a three-necked flask according to the literature [20]. Adjusting the pH value of the system was 6.0-7.0, and potassium persulfate/sodium hydrogen sulfite redox (0.03%–0.05%) were added. The solution was mixed and then bubbled with nitrogen for 30 min to displace dissolved oxygen then quickly capped. The polymerization was carried out at 50°C for 24 h. Products were precipitated and purified by ethanol to remove unreacted monomers and oligomers and then were baked in a vacuum drying oven at the 50°C to constant quality. The polymer molecular structure was shown in Figure 2, and characteristic parameters were shown in Table 1.

Table 1: Characteristic parameters of hydrophobically associating polyacrylamide (HAPAM)

Sample	Molecular weight	Degree of hydrolysis (%)	Critical association concentration (ppm)	Polydispersity index
HAPAM	6000000	25.0	800	1.46

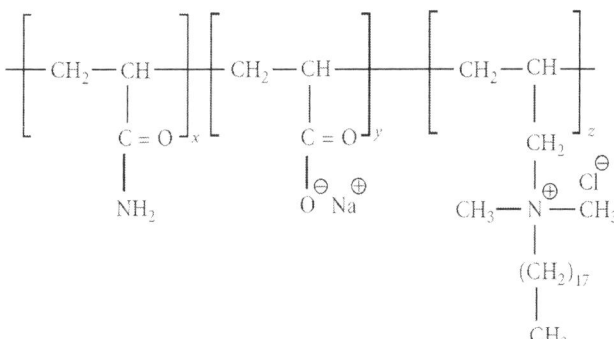

Figure 2: Molecular structure of hydrophobically associating polyacrylamide (HAPAM) (theoretical mole percents of , , and were about 79.7%, 20.2%, and 0.1%, resp.).

Solution Preparation

The aqueous solutions (5000 ppm) of surfactant and polymer were always freshly prepared using a mechanical stirrer for 8–10 h to form a consistent homogeneous solution at a low rotation per minute. Appropriate quantity of zwitterionic surfactants and hydrophobically associating polymers were dissolved carefully in distilled water for about 40 min [21, 22].

Measurement of Surface Tension and Interfacial Tension

In polymer-surfactant mixed systems, polymer concentrations were 500, 1000, 1500, and 2000 ppm, respectively, and BS-12 concentrations were 10–2000 ppm. The surfactant solution with or without HAPAM as outer-phase was injected into the glass tube, and 2 μL air or oil as inner-phase was put into the middle of the tube. Then the tube was enveloped with the plastic cover and put into the apparatus measuring the surface tension or interfacial tension at a temperature of 25°C [3]. In surface tension experiment, the outer-phase was polymer-surfactant mixed system and the inner-phase was air. Surface tensions of polymer-surfactant mixed systems were measured at a rotating velocity of 6000 rpm, when surface tension reached a minimum value and stopped the test. In interfacial tension (IFT) experiment, the outer-phase was polymer-surfactant mixed system and the inner-phase was dodecane. Interfacial tensions of mixed systems were measured at a rotating velocity of 5000 rpm, when interfacial tension reached a minimum value and stopped the test. In dynamic interfacial tension (DIFT) experiment, surfactant concentration was selected to make oil-water interfacial tension minimum in different polymer concentrations, and then interfacial tensions of the mixed system as time changing were measured a at a rotating velocity of 5000 rpm.

Measurement of Apparent Viscosity

In polymer-surfactant mixed systems, the polymer concentration was 1500 ppm and BS-12 concentrations were 10–2000 ppm. The viscosities of polymer-surfactant mixed systems were measured by Brookfield DV-III viscometer with a shearing rate of $7.34 \, s^{-1}$ and at a temperature of 25°C.

Laser Light Scattering Experiment

The diluted polymer-surfactant mixed system was dusted and filtered by the Millipore Corporation production of the disposable filter with a 0.8 μm aperture. The filtrate was collected in the sample pool. In the laser light scattering experiment, toluene as a standard solution, the laser wavelength was 532 nm and the measured temperature was 25°C [23, 24].

According to the light scattering theory [25–27], weight average molecular weight (M_w) and root mean square radius of gyration ($\langle R_g \rangle$) of polymers were measured by static light scattering. For high molecular weight polymers, the light scattering of polymer dilute solution can be expressed as:

$$\left(\frac{KC}{R_{vv}(q)}\right)^{1/2} \approx \left(\frac{1}{M_W}\right)^{1/2}\left(1+\frac{1}{6}\langle R_g^2 \rangle q^2\right)$$
$$\times (1+A_2 M_w C), \qquad (1)$$

where $K = 4\pi^2 n_0^2 (dn/dC)^2 /(N\lambda_0^4)$, K was associated with the constant of solvent nature and incident light frequency; n_0 was the refractive index of solution; C was the solution concentration, ppm; $R_{vv}(q)$ was the solvent effect of scattering intensity for different angles; λ_0 was the incident light wavelength; λ was the wavelength of incident light in solution, $\lambda=\lambda_0/n_0$; dn/dC was the refractive index increment and the ratio of solution refractive index and concentration, ppm; N was the Avogadro constant; $\langle R_g \rangle$ was root mean square radius of gyration and the chain quality centre to each chain segment average of squared distance. When the scattering angles were $\theta\rightarrow 0$ and concentrations were $C\rightarrow 0$, some parameters were obtained such as $\langle R_g \rangle$ and Mw by extrapolation [28, 29].

Hydrodynamic radius $\langle R_h \rangle$ of polymer molecules under different surfactant concentrations was measured by dynamic light scattering (DLS) [24]. In the dynamic light scattering, measurements were the light intensity-light intensity time related spectroscopy:

$$G^{(2)}(\tau) = A \cdot \left(1 + \beta \cdot \left|g^{(1)}(\tau)\right|^2\right),$$

(2)

where $G^{(2)}(\tau)$ was the autocorrelation function of light intensity; A was the baseline of the autocorrelation function; β was the experimental constant of constraint signal noise ratio, associated with the measuring experimental device; $g^{(1)}(\tau)$ was autocorrelation function of electric field. Its relationship with the line width distribution (Γ) is as follows:

$$g^{(1)}(\tau) = \int_0^\infty G(\Gamma) \exp(-\Gamma\tau) d\Gamma.$$

(3)

If the relaxation is caused entirely by diffusion, under the conditions of $C\rightarrow 0$ and $q\rightarrow 0$, $\Gamma = Dq^2$, D was the particle diffusion coefficient and q was the scattering vector. This moment, when the concentration was very low, D extrapolated to zero point and particle size distributions were obtained through Stokes-Einstein formulas $D = K_B T/(3\pi\eta d)$, where: K_B was Boltzmann constant; T was absolute temperature; η was solvent viscosity; d was particle diameter

Measurement of Molecular Aggregation Morphologies

AFM operating mode was tapping; Probe model was RTESP; operating frequency was 86 kHz; force constants were $1\sim 5$ Nm^{-1}. The system was stirred at a low velocity for 5 min to obtain a homogeneous solution concentration. For the AFM measurements, 0.1 mL of the prepared polymer-surfactant mixed system was dropped onto freshly cleaved mica, and the redundant solution was blown off by a stream of high purity nitrogen. Samples were measured by Nanoscope IIIa microscope in air at the ambient temperature [30].

RESULTS AND DISCUSSIONS

Surface Tension of Polymer-Surfactant Mixed System

It is well known that surfactants reduce the surface tension of water by getting adsorbed on the liquid-gas interface. The CMC, one of the main parameters for surfactants, is the concentration at which surfactant solutions begin to form micelles in large amounts [22]. Different techniques are used to examine the water-soluble polymer-surfactant aggregates formed in solution [31, 32]. Surface tensions of surfactant (BS-12) solutions at different concentrations were measured and plotted as a function of concentrations (Figure 3). Figure 3 shows that the CMC and the surface tension have been a large change after adding polymers into surfactant solutions. When polymers are not added, the CMC and the CMC of the surface tension are minimal. The CMC value of BS-12 is about 300 ppm and the surface tension is about 31.3 mN/m. After the polymers being added, the surfactant CMC and the CMC of surface tensions gradually increase with the increment of polymer concentrations. When HAPAM concentration is about 2000 ppm, the CMC of the mixed system is about 500 ppm and the surface tension increases to 33.4 mN/m. Khan et al. also found the same behavior of polyacrylamide solutions in the presence of SDBS [22]. The reason for this phenomenon is that hydrophobic groups of polymers will interact with surfactant hydrophobic parts, and some surfactants are shackled in the bulk phase. Shackled surfactants increase with the increment of polymer concentrations. They need to consume more surfactants to form micelles [3]. Therefore, reaching critical micelles requires higher surfactant concentrations. Mixed micelles of polymers/surfactants that are formed have been inhibitory effect to the surface tension of surfactants in solution.

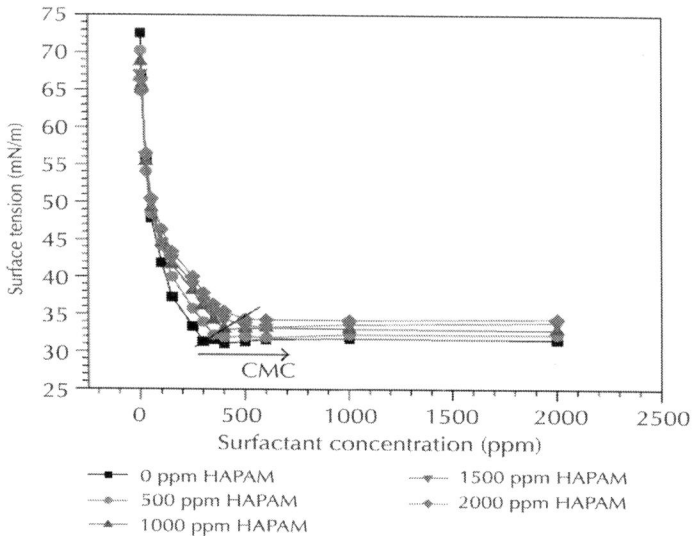

Figure 3: Effect of HAPAM concentrations on surface tensions of BS-12.

Interfacial Tension of Polymer-Surfactant Mixed System

The polymer-surfactant mixed system has been applied in the oil field owing to reducing the mobility ratio decrease interfacial tension (IFT) between the water and the oil [33, 34]. It has been reported that addition of polymers increases the IFT of ionic surfactants [3, 35]. Interfacial tensions of surfactant (BS-12) solutions at different concentrations are measured, and results are shown in Figure 4. Figure 4 shows that oil-water IFT is decreased with the increment of surfactant concentrations. After a turning point, the IFT becomes to balance. If there are no polymers, the turning point of surfactant concentration and the IFT is minimal. The turning point of surfactant concentration is about 300 ppm and the IFT is about 0.29 mN/m. With the increment of polymer concentrations, the turning point of surfactant concentration and the IFT have shown a trend of increment. This phenomenon is similar to Figure 2. Hydrophobic groups of polymers will interact with surfactant hydrophobic parts and bound part surfactants in the bulk phase. Ultimately, the ability of reduced oil-water IFT is declined. The shackled effect is enhanced with the polymer concentrations increasing. Therefore the reduced oil-water IFT needs to consume more surfactants. On the other hand, the IFT of surfactants can be inhibited by the viscosity of polymer solutions. Chen et al. also confirm the result [3].

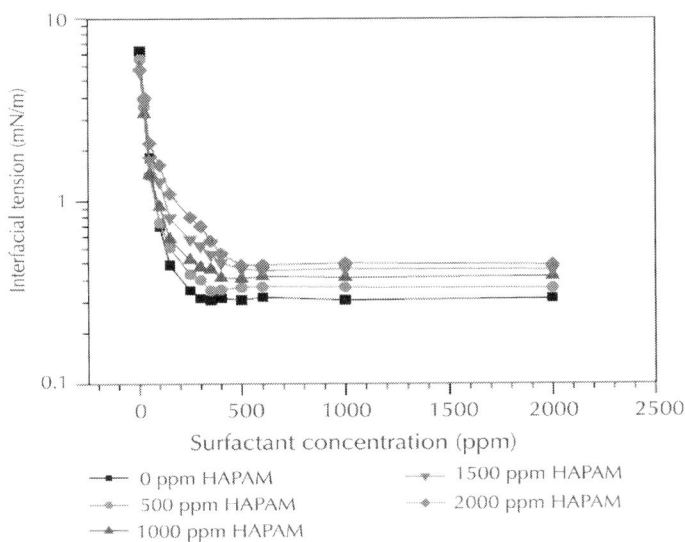

Figure 4: Effect of HAPAM concentrations on interfacial tensions of BS-12.

Figure 5 shows the result that the oil-water dynamic interfacial tension (DIFT) of BS-12 is affected by the HAPAM concentrations. The DIFT is changed with time increasing. The reduced surfactant IFT and reaching steady state that required time can be reflected by DIFT characteristics. When surfactant concentrations are 500 ppm, the DIFT change is faster at lower polymer concentrations. Therefore oil-water IFT reaching a minimum need less time. The higher concentrations the polymer is, the longer the time reaching a minimum of IFT is. The reason is that surfactants and associating polymers have the strong interaction and oil-water interface diffusion rates of surfactants which are significantly affected. In addition, the system viscosity increases with the increment of polymer concentrations, and the surfactant spread speed is also slowed. As a result, the IFT reaching a minimum needs longer times at the higher polymer concentrations.

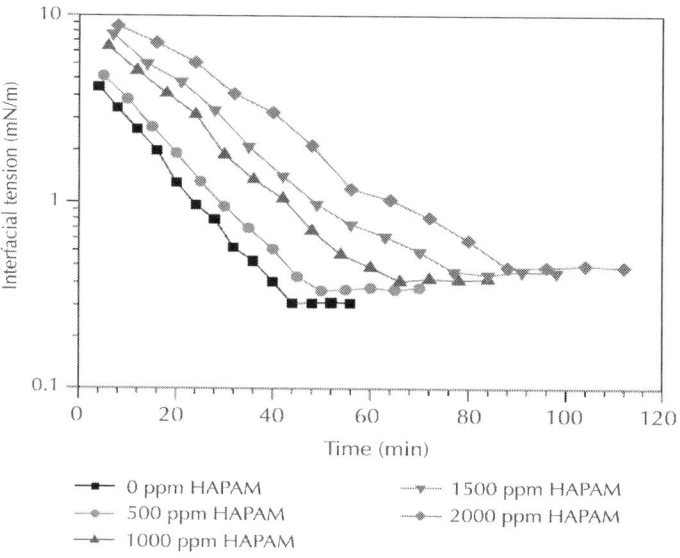

Figure 5: Effect of HAPAM concentrations on the dynamic interfacial tensions of BS-12.

Apparent Viscosity of Polymer-Surfactant Mixed System

The effect of BS-12 concentrations on the viscosity of polymers is shown in Figure 6. Figure 6 shows that the system viscosity increases first and then declines with the increment of BS-12 concentrations. When the BS-12 concentration is about 100 ppm, the system viscosity is the largest. Badoga et al. [36] and Jiang et al. [37] have reported that the viscosity of polymer-surfactant mixed systems increases first and then declines with the increment of surfactant concentrations. When the addition of surfactant concentrations is lower, surfactant molecules in single molecule state are distributed in aqueous solution. Surfactant molecular ions and HAPAM molecular chains interacting with each other make the interaction of hydrophobic groups forming inner salt keys be opened, and the intermolecular association is formed. At this moment, the association between polymer molecules promoted role and polymer molecular chains is more diastolic because of the addition of surfactants. The solution viscosity is increased with the increment of surfactant concentrations. When surfactant concentrations exceed a certain value, the interaction between surfactants and hydrophobic groups of polymer chain segments is further enhanced. Mixed micelles of polymer-surfactant are formed. On the other hand, the intermolecular association of polymers is shielded with the increase of surfactant micellar numbers; thus polymer network structures are damaged and collapsed. The viscosity of mixed systems is decreased.

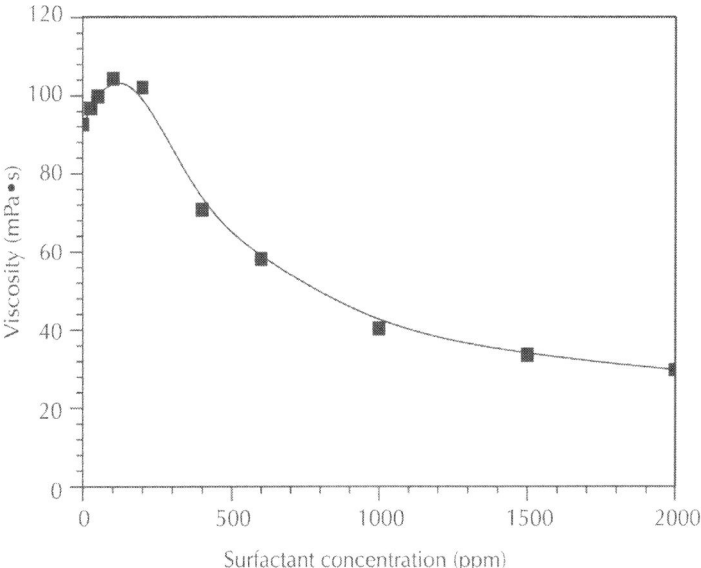

Figure 6: Effect of BS-12 concentrations on the viscosity of 1500 ppm HAPAM.

Laser Light Scattering Experiments

It has been reported [23, 38] that addition of ionic surfactants influences the molecular structure of the polymer. The effects of BS-12 concentrations on apparent weight average molecular weights (M_w) of polymers are tested at a 25°C. Their result is shown in Figure 7. The root mean square radius of gyration ($\langle R_g \rangle$) is characteristic parameters of the polymer and directly reflects the conformation of polymer chains. In order to reduce the effect of polymer concentrations, the preparation concentration of polymers is 2 ppm in experiment. For the dilute polymer solution, $\langle R_g \rangle$ can be concluded through extrapolation of the same concentrations with different angles in solution; then some $\langle R_g \rangle$ are received under different surfactant concentrations in the same way. Their results are shown in Figure 8.

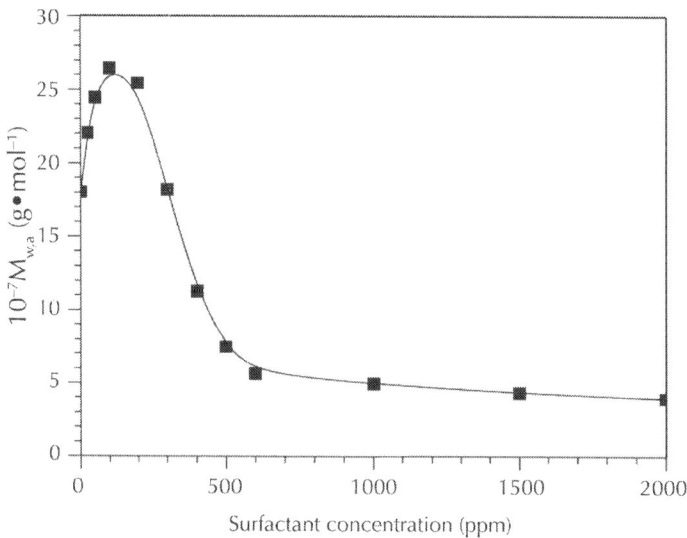

Figure 7: Effect of BS-12 concentrations on apparent weight average molecular weight (M_w) of HAPAM.

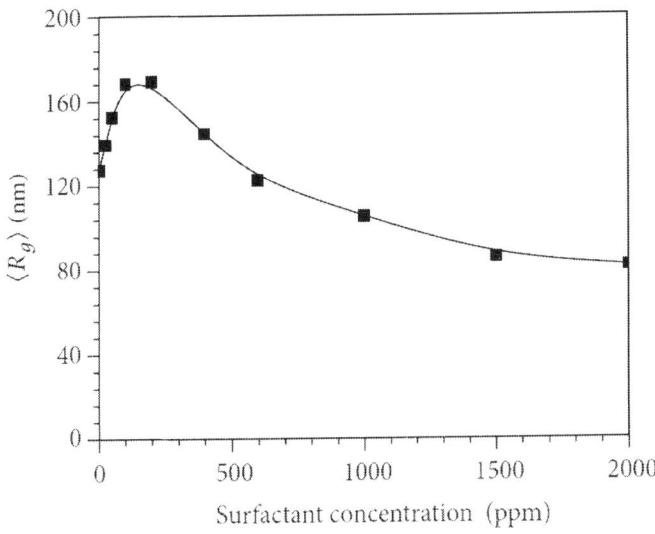

Figure 8: Effect of BS-12 concentrations on root mean square radius of gyration $\langle R_g \rangle$ of HAPAM.

Figures 7 and 8 show that M_w and $\langle R_g \rangle$ of HAPAM increase first and then decline with the increment of BS-12 concentrations. When there are a few surfactants in the polymer-surfactant mixed system, the surfactant molecules interacting with polymer hydrophobic groups, intermolecular association of polymers is strengthened and polymers are more likely to gather to form super molecular structures. M_w and $\langle R_g \rangle$ show a trend of increment. When surfactant concentrations are about 100 ppm, the system viscosity is the largest. With the further increment of surfactant concentrations, hydrophobic groups of polymer molecules are inhibited by cationic groups of surfactant molecular chains. The intramolecular association of associating polymers forms inner salt key. M_w and $\langle R_g \rangle$ become smaller. When surfactant concentrations are more than its critical micelle concentrations, the number of surfactant micelles is increased. Hydrophobic groups of associating polymers are separated by surfactant micelles. The intermolecular association is weakened and supramolecular aggregations are dismantled. Thus, M_w and $\langle R_g \rangle$ are further smaller.

In order to study zwitterionic surfactant (BS-12) effect on hydrodynamic sizes of polymers in solution, the preparation concentration of polymers is 2 ppm in the experiment. Particle size distributions and hydrodynamic radius ($\langle R_h \rangle$) of the polymer under different surfactant concentrations are measured by dynamic light scattering at a 25°C and the scattering angle is 90°. Their results are shown in Figures 9 and 10. Figure 9 shows that particle sizes of HAPAM are unimodal distribution under no surfactant condition. When surfactant concentrations are about 100 ppm, the particle size distribution of HAPAM is a wider unimodal and moves to the right. Surfactants can enhance intermolecular association of polymers and make polymer chains stretch and hydrodynamic radius increase. When surfactant concentrations are more than 400 ppm, particle sizes of HAPAM are multimodal distributions and wider unimodal move to left. The reason is that surfactant micelles are increased with the increment of surfactant concentrations in solution. Some polymer hydrophobic groups are embedded by surfactant micelles. The intermolecular association of HAPAM is partially blocked, and the hydrodynamic radius appears to be reducing.

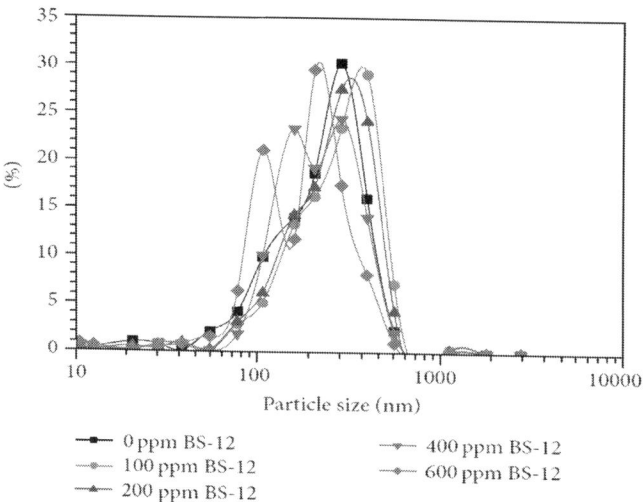

Figure 9: Particle size distributions of HAPAM with different BS-12 concentrations.

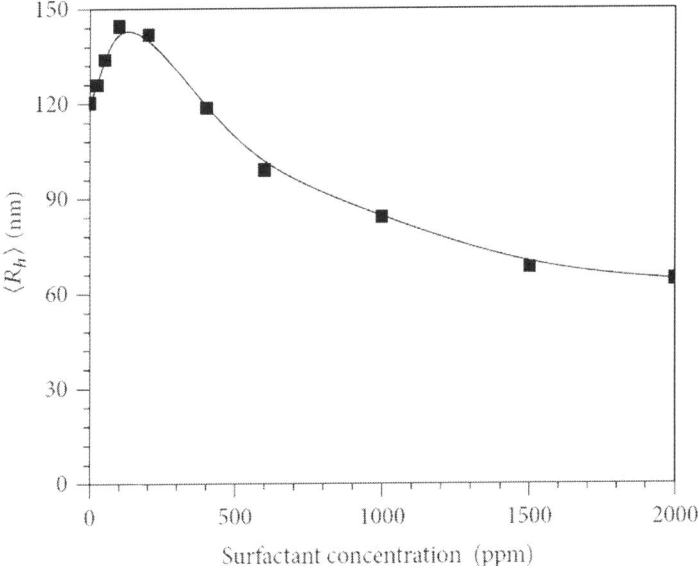

Figure 10: Effect of BS-12 concentrations on the hydrodynamic radius ($\langle R_h \rangle$) of HAPAM.

Figure 10 shows that $\langle R_h \rangle$ is increased because of a small amount of surfactants to be added. When surfactant concentrations are lower, the intermolecular association of HAPAM is strengthened, and $\langle R_h \rangle$ is increased. But the surfactant concentrations exceed a certain value; polymer aggregations are dismantled; therefore, ($\langle R_h \rangle$) is reduced.

Molecular Aggregation Morphologies of Polymer-Surfactant Mixed System

The previous research results had been confirmed that space network structures of hydrophobically associating polymer are formed exceeding the critical association concentration (CAC) of polymer [3, 24]. When ionic surfactants are added, molecular aggregation morphologies of HAPAM are affected. Different BS-12 concentrations affecting molecular aggregation morphologies are observed. Their results are shown in Figure11. Figure 11(a) is an AFM photo of HAPAM without surfactants. When the HAPAM concentration exceeds the CAC, it can be formed obvious spatial network structures in distilled water [39]. Figure 11(b) is an AFM photo of HAPAM solution to add 100 ppm surfactants. Compared with Figures 11(a) and 11(b), when BS-12 concentrations are about 100 ppm, the space network structures become more intense in solution, and the connecting mesh chain beams are thicker, especially the intersection part of chain beam. Figure 11(c) is an AFM photo of HAPAM solution to add 400 ppm surfactants. Compared with Figures 11(b) and 11(c), when BS-12 concentrations are about 400 ppm, the space network structures become sparser. When surfactants continue to be added, the connecting network chain beams are thinner. These results show that adding a few surfactants has a promoting effect on the self-assembly of polymer molecules, but the self-assembly of polymer molecules is inhibited for adding too many surfactants.

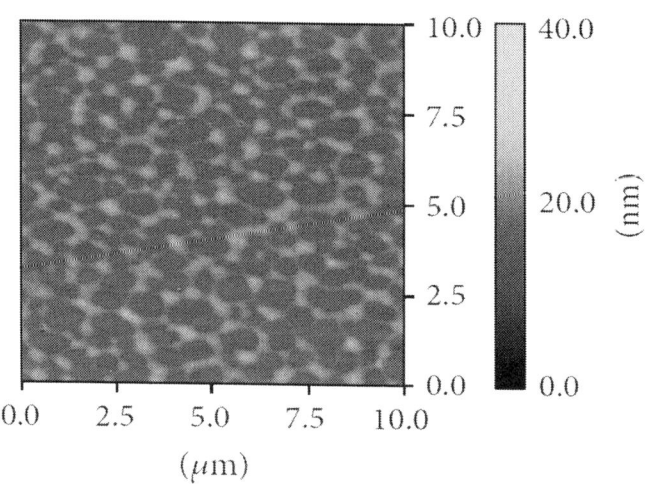

(a)

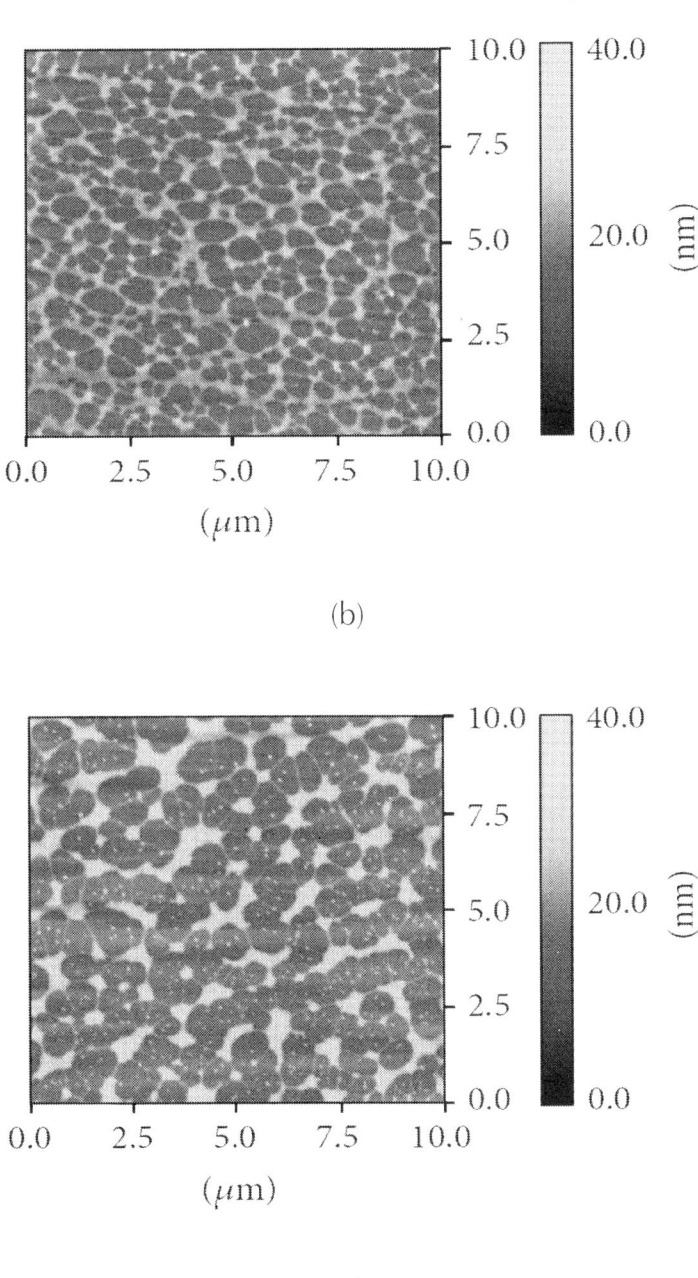

Figure 11: AFM images of 1500 ppm HAPAM with different BS-12 concentrations: (a) 0 ppm BS-12, (b) 100 ppm BS-12, and (c) 400 ppm BS-12.

CONCLUSIONS

- Adding polymers into dodecyl dimethyl betaine (BS-12) solutions, the CMC and surface tensions of the CMC are increased with the increment of HAPAM concentrations.
- Zwitterionic surfactant (BS-12) can reduce dodecanoic-water interfacial tension to about 0.3 mN/m. BS-12 has been a good ability to reduce the oil-water interfacial tension. The surfactant (BS-12) interface activity is affected by HAPAM. The required time of interfacial tension balance is longer with polymer concentrations increasing.
- When the surfactant (BS-12) concentrations are lower, the apparent viscosity, apparent weight average molecular weights (M_w), root mean square radius of gyration ($\langle Rg \rangle$), and hydrodynamic radius ($\langle R_h \rangle$) of HAPAM increase with the increment of BS-12 concentrations. When surfactant concentrations are 100 ppm, they are maximum; surfactant concentrations continue to increase and they begin to decline. BS-12 has a great influence on performances of HAPAM solutions.
- The hydrophobically associating polymer (HAPAM) can form the obvious spatial network structures exceeding the critical association concentration (CAC) in distilled water. When added surfactant (BS-12) concentrations are about 100 ppm, the space network structures become more intense, and the connecting network chain beams are thicker. BS-12 concentrations continue to increase, when concentrations are about 400 ppm, loose network structures are formed, and partially loose network structures are broken.

ACKNOWLEDGMENTS

The authors gratefully appreciated the National Science and Technology Major Projects, China (no. 2011ZX05011). The authors appreciated the State Key Laboratory of Oil and Gas Reservoir Geology and Exploitation for experiment help too.

REFERENCES

1. C. Wang, X.-R. Li, and P.-Z. Li, "Study on preparation and solution properties of hydrophobically associating polyacrylamide by emulsifier-free ultrasonic assisted radical polymerization," Journal of Polymer Research, vol. 19, no. 8, pp. 9933–9939, 2012.

2. S.-L. Cram, H.-R. Brown, M.-S. Geoffrey, D. Hourdet, and C. Creton, "Hydrophobically modified dimethylacrylamide synthesis and rheological behavior," Macromolecules, vol. 38, no. 7, pp. 2981–2989, 2005.

3. H. Chen, E.-X. Li, Z.-B. Ye, L.-J. Han, and P.-Y. Luo, "Interaction of hydrophobically associating polyacrylamide with gemini surfactant," Acta Physico-Chimica Sinica, vol. 27, no. 3, pp. 671–676, 2011.

4. G.-O. Yahaya, A.-A. Ahdab, S.-A. Ali, B.-F. Abu-Sharkh, and E.-Z. Hamad, "Solution behavior of hydrophobically associating water-soluble block copolymers of acrylamide and N-benzylacrylamide,"Polymer, vol. 42, no. 8, pp. 3363–3372, 2001.

5. F. S. Hwang and T. E. Hogen-Esch, "Effects of water-soluble spacers on the hydrophobic association of fluorocarbon-modified poly(acrylamide)," Macromolecules, vol. 28, no. 9, pp. 3328–3335, 1995.

6. M. Li, M. Jiang, Y.-X. Zhang, and Q. Fang, "Fluorescence studies of hydrophobic association of fluorocarbon-modified poly(N-isopropylacrylamide)," Macromolecules, vol. 30, no. 3, pp. 470–478, 1997.

7. Y.-J. Feng, L. Billon, B. Grassl, G. Bastiat, O. Borisov, and J. François, "Hydrophobically associating polyacrylamides and their partially hydrolyzed derivatives prepared by post-modification. 2. Properties of non-hydrolyzed polymers in

pure water and brine," Polymer, vol. 46, no. 22, pp. 9283–9295, 2005.

8. P. Deo and P. Somasundaran, "Interactions of hydrophobically modified polyelectrolytes with nonionic surfactants," Langmuir, vol. 21, no. 9, pp. 3950–3956, 2005.

9. G. Nizri, S. Lagerge, A. Kamyshny, D. T. Major, and S. Magdassi, "Polymer-surfactant interactions: binding mechanism of sodium dodecyl sulfate to poly(diallyldimethylammonium chloride)," Journal of Colloid and Interface Science, vol. 320, no. 1, pp. 74–81, 2008.

10. A.-E. Goddard, M.-L. Francisco, M.-J. Arturo, and H.-A. Roque, "Two-dimensional colloidal aggregation: concentration effects," Journal of Colloid and Interface Science, vol. 246, no. 2, pp. 227–234, 2002.

11. D.-X. Wang, L. Luo, L. Zhang, Y.-Y. Wang, S. Zhao, and J.-Y. Yu, "Study on interfacial interaction between hydrophobically modified polyacrylamide and surfactants," Acta Physico-Chimica Sinica, vol. 21, no. 11, pp. 1205–1210, 2005.

12. A.-S. Anna, R.-A. Campbell, and C.-D. Bain, "Dynamic adsorption of weakly interacting polymer/surfactant mixtures at the air/water interface," Langmuir, vol. 28, no. 34, pp. 12479–12492, 2012.

13. N. Beheshti, A.-L. Kjøniksen, K. Zhu, K. D. Knudsen, and B. Nyström, "Viscosification in polymer-surfactant mixtures at low temperatures," Journal of Physical Chemistry B, vol. 114, no. 19, pp. 6273–6280, 2010.

14. L.-D. Jiang, B.-J. Gao, and L. Gang, "Interaction between cationic Gemini surfactant with hydrophobically associatied polyacrylamide of a new family," Acta Physico-Chimica Sinica, vol. 23, no. 3, pp. 337–342, 2007.

15. N.-V. Sastry and H. Hoffmann, "Interaction of amphiphilic block copolymer micelles with surfactants,"Colloids and Surfaces A, vol. 250, no. 1–3, pp. 247–261, 2004.

16. L. Piculell, M. Egermayer, and J. Sjöström, "Rheology of mixed solutions of an associating polymer with a surfactant.

Why are different surfactants different?" Langmuir, vol. 19, no. 9, pp. 3643–3649, 2003.
17. G.-L. Smith and C.-L. McCormick, "Water-soluble polymers. 79-Interaction of microblocky twin-tailed acrylamido terpolymers with anionic, cationic, and nonionic surfactants," Langmuir, vol. 17, no. 5, pp. 1719–1725, 2001.
18. J.-R. Enrique, S. Joseph, and C. Francoise, "Effect of surfactant on the viscoelastic behavior of semidilute solutions of multisticker associating polyacrylamides," Langmuir, vol. 16, no. 23, pp. 8611–8621, 2000.
19. X.-Y. Wang, Y.-J. Li, J.-B. Wang et al., et al., "Interactions of cationic gemini surfactants with hydrophobically modified poly(acrylamides) studied by fluorescence and microcalorimetry," Journal of Physical Chemistry B, vol. 109, no. 26, pp. 12850–12855, 2005.
20. H. Chen, W.-T. Lu, Z.-B. Ye, L.-J. Han, and P.-Y. Luo, "Influence of hydrolysis degree on properties of associating polymers solution," Oilfield Chemistry, vol. 29, no. 2, pp. 190–194, 2012.
21. E. Minatti and D. Zanette, "Salt effects on the interaction of poly(ethylene oxide) and sodium dodecyl sulfate measured by conductivity," Colloids and Surfaces A, vol. 113, no. 3, pp. 237–246, 1996.
22. M. Y. Khan, A. Samanta, K. Ojha, and A. Mandal, "Interaction between aqueous solutions of polymer and surfactant and its effect on physicochemical properties," Asia-Pacific Journal of Chemical Engineering, vol. 3, no. 5, pp. 579–585, 2008.
23. L.-J. Han, Z.-B. Ye, H. Chen, and P.-Y. Luo, "Self-assembly of hydrophobically associating polyacrylamide and gemini surfactant," Acta Physico-Chimica Sinica, vol. 28, no. 6, pp. 1405–1410, 2012.
24. H. Chen, X.-Y. Wu, Z.-B. YE, L.-J. Han, and P.-Y. Luo, "Self-assembly behavior of hydrophobically associating polyacrylamide in salt solution," Acta Physico-Chimica Sinica, vol. 28, no. 4, pp. 903–908, 2012.

25. Q.-W. Zhang, J. Ye, Y.-J. Lu et al., "Synthesis, folding, and association of long multiblock (PEO 23-b-PNIPAM124)750 chains in aqueous solutions," Macromolecules, vol. 41, no. 6, pp. 2228–2234, 2008.
26. L. Hong, F.-M. Zhu, J.-F. Li, T. Ngai, Z.-W. Xie, and C. Wu, "Folding of long multiblock copolymer (PI-b-PS-b-PI)n chains prepared by the Self-Assembly Assisted Polypolymerization (SAAP) in cyclohexane," Macromolecules, vol. 41, no. 6, pp. 2219–2227, 2008.
27. D. Xie, X. Ye, Y.-W. Ding et al., "Multistep thermosensitivity of Poly(N-n-propylacrylamide)-block-poly(N-isopropylacrylamide)-block-poly(N,N-ethylmethylacrylamide) triblock terpolymers in aqueous solutions as studied by static and dynamic light scattering," Macromolecules, vol. 42, no. 7, pp. 2715–2720, 2009.
28. X. Wang, X. Qiu, and C. Wu, "Comparison of the coil-to-globule and the globule-to-coil transitions of a single poly(N-isopropylacrylamide) homopolymer chain in water," Macromolecules, vol. 31, no. 9, pp. 2972–2976, 1998.
29. P.-A. Fuierer, B. Li, and H. S. Jeon, "Characterization of particle size and shape in an ageing bismuth titanate sol using dynamic and static light scattering," Journal of Sol-Gel Science and Technology, vol. 27, no. 2, pp. 185–192, 2003.
30. R. Zhang, Z.-B. Ye, L. Peng, N. Qin, Z. Shu, and P.-Y. Luo, "The shearing effect on hydrophobically associative water-soluble polymer and partially hydrolyzed polyacrylamide passing through wellbore simulation device," Journal of Applied Polymer Science, vol. 127, no. 1, pp. 682–689, 2012.
31. Y. Dong and D.-C. Sundberg, "Estimation of polymer/water interfacial tensions: hydrophobic homopolymer/water interfaces," Journal of Colloid and Interface Science, vol. 258, no. 1, pp. 97–101, 2003.
32. M. Nedjhioui, N. Moulai-Mostefa, A. Morsli, and A. Bensmaili, "Combined effects of polymer/surfactant/oil/alkali on physical chemical properties," Desalination, vol. 185, no. 1–3, pp. 543–550, 2005.

33. J.-X. Liu, Y.-J. Guo, J. Hu, et al., "Displacement characters of combination flooding systems consisting of gemini-nonionic mixed surfactant and hydrophobically associating polyacrylamide for bohai offshore oilfield," Energy Fuels, vol. 26, no. 5, pp. 2858–2864, 2012.
34. Y.-J. Guo, J.-X. Liu, X.-M. Zhang et al., "Solution property investigation of combination flooding systems consisting of gemini-non-ionic mixed surfactant and hydrophobically associating polyacrylamide for enhanced oil recovery," Energy and Fuels, vol. 26, no. 4, pp. 2116–2123, 2012.
35. H.-J. Gong, X. Xin, G. Y. Xu, and Y.-J. Wang, "The dynamic interfacial tension between HPAM/$C_{17}H_{33}COONa$ mixed solution and crude oil in the presence of sodium halide," Colloids and Surfaces A, vol. 317, no. 1–3, pp. 522–527, 2008.
36. S. Badoga, S.-K. Pattanayek, A. Kumar, and L.-M. Pandey, "Effect of polymer-surfactant structure on its solution viscosity," Asia-Pacific Journal of Chemical Engineering, vol. 6, no. 1, pp. 78–84, 2011.
37. L.-D. Jiang, B.-J. Gao, and L. Gang, "Interaction between cationic Gemini surfactant with hydrophobically associatied polyacrylamide of a new family," Acta Physico-Chimica Sinica, vol. 23, no. 3, pp. 337–342, 2007.
38. Y.-J. Mei, Y.-X. Han, H. Zhou, L. Yao, and B. Jiang, "Synergism between hydrophobically modified polyacrylic acid and wormlike micelles," Acta Physico-Chimica Sinica, vol. 28, no. 7, pp. 1751–1756, 2012.
39. R. Zhang, Z.-B. Ye, and P.-Y. Luo, "The atomic force microscopy study on the microstructure of the polymer solution," Journal of Chinese Electron Microscopy Society, vol. 29, no. 5, pp. 475–481, 2010.

Chapter 7

Opportunities and Challenges pf Robotics and Automation on Offshore Oil & Gas Industry

Heping Chen[1], Samuel Stavinoha[1], Michael Walker[1], BiaoZhang[2], and Thomas Fuhlbrigge[2]

[1]Ingram School of Engineering, Texas State University, San Marcos, USA
[2]ABB Corporate Research Center, Windsor, USA

ABSTRACT

The oil and gas industry will continue to boom in the coming few decades. Obtaining oil and gas from conventional and non-conventional resources will become more and more challenging.

This intensifying need will impose very considerable demands on work force, financial and technology capabilities. Since the future supplies of oil and gas are to expand, advanced technology will become increasingly necessary to obtain access to more challenging conventional and non-conventional resources. Therefore oil and gas technologies will be very costly to operate in the coming future due to hostile, hard-to-reach environments. The offshore oil industry will become a complicated myriad of advanced equipment, structures, and work force. Our objectives are to identify potential applications and research directions of robotics and automation in the oil & gas field and explore the obstacles and challenges of robotic and automation applications to this area. This study performs the necessary survey and investigation about the work conditions of robotics and automation equipment in the oil and gas industry, especially offshore oil rigs. The oil & gas industry processes are first investigated. The personals and tasks are then explored. Furthermore, this paper reviews the current robotic automation technology. The challenges and requirements are identified for robotics and automation equipment in the oil and gas industry. The requirements of robotics and automation in the oil & gas industry are presented. Future research opportunities are discussed from a technical perspective.

INTRODUCTION

The oil and gas industry will continue to boom in the coming few decades. The oil and gas demand will grow rapidly in the next two decades [1]

The intensifying need to obtain oil and gas from more hostile, hard-to-reach environments will increase the operation cost rapidly in the coming future. Hence, the oil and gas industry keeps looking for lower-cost solutions.

To be competitive and to improve their profit margins, oil & gas companies are committed to cost reduction. They also look for ways to minimize employee costs and improve manufacturing efficiencies and quality besides seeking lower-cost suppliers and less-expensive raw materials. Because of the rising cost of employee salary and benefits like health care, the cost reduction effort in oil & gas companies is offset. Also high employee turnover adds the costs of retraining. Therefore, the oil and gas companies are looking for new technologies to reduce the labor cost. Also safety is a big concern in the oil and gas production. Using robotics in inspection, maintenance and repair could greatly improve the safety and efficiency.

The oil & gas industry's presence is evident in its global networks of market supply and demand relationships. When there are fluctuations, regardless of their origins, consumers are affected in all over the world. Prices respond to changing markets with upward volatility because of an inelastic demand for oil and petroleum products. The period of high oil prices from 2004 through 2008 led to a steady demand for petroleum. As this market trend persists, oil & gas companies have a window of opportunity to maximize efficiency and productivity to moderate the petroleum market.

One solution to both the need for efficiency and maximum production and the capabilities required to further exploration is to implement robotics and automation in offshore oil & gas environments. Because the offshore oil & gas processes require advanced technologies, offshore environments will deploy the safest, most secure and consistent operations by utilizing industrial robotics and automation, and the latest software and mechanical devices.

In order to investigate the challenges of robotics and automation in oil and gas industry, the necessary survey and investigation about the oil & gas industry processes, the personals and tasks should be explored first. The work conditions must be discussed to explore the requirements of robotics and automation equipment in the oil and gas industry, especially on offshore rigs. To meet the requirements and develop robotics and automation equipment in such work conditions, this paper reviews the current technology that has been developed and discusses the future research opportunities in the oil and gas industry.

OIL AND GAS INDUSTRIAL PROCESSES

The offshore oil & gas industry is a complicated myriad of advanced equipment, structures, and work force. With a proper knowledge of offshore oil and gas rig environments, the applications of industrial robotics and automation are less abstract. Before any real vision of the potential roles robotics and automation in offshore oil processes can emerge, those processes must be enumerated appropriately.

There are many products and services related to oil and gas with an equally substantial potential for markets within the industry. There are three stages through which petroleum products pass: upstream, downstream and midstream models.

The upstream oil sector commonly refers to the searching for, recovery and production of crude oil and natural gas. It is also known as the exploration and production (E & P) sector, including searching for potential oil and gas fields, drilling of exploratory wells, and subsequently operating the wells to recover and bring the crude oil and/or natural gas to the surface [2].

The midstream oil and gas sector is the relay point for the upstream sector's products. Midstream processes commonly refer to processing, transport, and storage of these products [3]. Because it is possible to produce pipeline quality gas for direct sale to an interstate or intrastate natural gas pipeline in the midstream sector, some treatment or processing of natural gas may occur in the midstream sector and bypass the downstream oil and gas sector completely. The midstream typically links the supply of the oil industry to the demand for energy commodities [3].

The downstream oil sector refers to the refining of crude oil and the selling and distribution of natural gas, as well as other products derived from crude oil such as liquefied petroleum gas (LPG), gasoline or petrol, jet fuel, diesel oil, other fuel oils, asphalt and petroleum coke [4] [5]. The downstream industry touches consumers through thousands of products. These products include petrol, diesel, jet fuel, heating oil, natural gas and propane to asphalt, lubricants, synthetic rubber, plastics, fertilizers, antifreeze, pesticides, and pharmaceuticals [4] [5] .

This paper provides a fundamental knowledge base to utilize in understanding the environment in which offshore industrial robotics and automation would operate, as well as cover the challenges that exist within that environment. The processes described here relate primarily to the upstream oil and gas sector, as that particular sector hosts the majority of opportunities for robotics and automation. The oil and gas processes and the three sector model they fall into tend to parallel across the onshore and offshore industries, however the processes will be distinguished when necessary if a distinct observation is being made.

The major oil and gas extraction processes include the materials and equipment used and the processes employed. There are four major processes in the oil and gas extraction industry: (1) exploration, (2) well development, (3) production, and (4) site abandonment [6] . After these processes are completed, the production process enters. It is likely the process in which robotics and automation have the largest potential to increase efficiency and create a safer environment for offshore oil and gas rigs, all while cutting construction costs for human necessitated rig designs [6] .

After the Deep Water Horizon oil spill in 2010, the Bureau of Ocean Energy Management, Regulation and Enforcement (BOEMRE) has implemented new mandatory regulations to replace old protocols for the offshore oil & gas industry. Consequences of these new regulations are evident in the day-to-day operations in the oil & gas industry, especially regarding inspection and maintenance of equipment. This is one area of operation that robotics and automation can dramatically improve efficiency, precision, safety, and decrease costs to companies. It is no stretch of imagination to suggest that robots and automation will soon be the primary means to effectively satisfy many of these new regulations.

Challenges and Requirements

In this section, the environments that the robotics and automation will be deployed are investigated. The hardware and software requirements are discussed.

Challenges in Oil and Gas

The deep waters of the Gulf of Mexico, the frigid regions of Russia, and the hot, dusty, undeveloped deserts of the Middle East are merely the geographic challenges facing today's oil and gas exploration and production industry [7] . The work conditions on offshore installations are the first thing to look at when analyzing the environments. The most important ones are as follows:

- Atmosphere: The atmospheric conditions on offshore platforms are quite unfriendly. Due to the substances used and generated during the processing of hydro carbon resources, the following three types of gases can occur separately and combined: explosive, toxic and corrosive.
- Unsheltered maritime environment: Except for the living quarters and a few technical rooms offshore platforms are partially sheltered and unsheltered. This means there is no sufficient protection against saltwater spray and direct sun light which is also reflected from the sea surface.
- Heavy weather: Wind with high speed and squalls, rain, hail and snow. All these weather conditions occur more often and more intense offshore than onshore.
- Extreme ambient temperature: Depending on the region the platform is located there can be extreme high and low temperatures. Humidity is also ranging from lower values up to condensing.
- Constraint space and/or walkways: The width of typical walkways is about 0.7 - 0.75 m.
- Offshore rigs have further logistical issues: (a) it is highly expensive to have people working on the rig as they must

be housed and protected; (b) in the case of emergency, it must be possible to evacuate personnel quickly. As oil and gas exploration pushes into more hostile and remote regions, these difficulties become serious obstacles to the financial viability of an offshore installation.

With a basic comprehension of the processes and the personnel tasks involved in the offshore oil and gas industry, the lingering factor is the role that robotics and automated systems could potentially fill. While this is a relatively young market, there are a few examples of robotics and automation used for oil processes, but only a few. Reason for such hesitation derives from the logistical challenges that come with the implementation of a robot or an automated system in an offshore environment, as well as from a general lack of prototypes on the market.

The disaster at the Macondo Prospect, which is called the Deepwater Horizon oil spill [8] , has spawned a great deal of movement and some legislation. Much of the legislation requires an increase in inspection and maintenance at offshore rigs, where robotics and industrial automation will find a role in enabling those in the industry to meet these new standards.

To generate application scenarios for mobile robots, the operations carried out on these types of platforms must be understood. There are scheduled and occasional operations. The scheduled operations are tasks planed in the daily operation schedule. The occasional operations are those triggered by external influence on a more or less random basis. The most important scheduled operations are:

- Inspection: gauge readings and valve and lever position readings
- Monitoring: gas level, check for leakage, acoustic anomalies, surface condition and check for intruders.
- Maintenance: gas and fire detector test, sampling, pigging, cleaning, refilling and pipelines.
- The most frequent occasional operations are:

- Valve and lever operation: change pressure, change flow rate and start or stop equipment operation.
- Gas leakage: identify and locate, stop dangerous operations (welding, cutting, ...) and secure area and stop leakage and monitor concentration drop.
- Fire: identify and locate fire.

The robot will have to operate at various levels of automation: fully automatic, semi-automatic and manual. Fully automatic operations require no human intervention. There will be various tasks using semi-automatic operations, which will require varying degrees of human interaction. This is quite different from the more traditional industrial robotic applications because human decision makers must be within the control loop to collaborate with the robots and the control system. The successful application of robotics and automation in the oil and gas industry will rely on the seamless integration of man, technology, and organization. Compared to the fully automated offshore robots, the inspection robots are the simplest because they may need constant human involvement. The manipulation robot is more complex than the inspection robot because it has to make decisions while performing different tasks. In the short term, the inspection robot could have many applications in the oil & gas industry.

Requirements for Hardware Development

In order to be suitable for dependable and useful offshore operations, the following basic requirements must be met by the hardware of the mobile inspection and manipulation robots:

- The robot must be certified—or as a prototype be certifiable. The robotic system must be explosive-proof, weather-proof and salt-water-proof [9].
- The electronics must be suitable for harsh environments.
- The drive systems of the robot must be suitable for the hostile environments and the very special floor conditions.

- The robot has to maneuver in confined spaces. Therefore its size must be adapted to the previously defined reference passage.
- The robot must be equipped with highly reliable sensors to perceive its surroundings, especially to detect obstacles.
- If the robot is supposed to navigate in different levels, the robot must be able to move vertically on simple ladder-type steel profiles.
- The robot must be equipped with sensors to track its position for autonomous motion.
- The robot must be equipped with a manipulator to handle objects and position sensors.
- The robot must have appropriate application sensors and tools to execute inspection and manipulation tasks autonomously, semiautonomously or manually.
- The robot must be able to communicate with a central control station, for example by Wireless LAN.

Requirements for Communication

A major challenge for teleoperation within the oil and gas industry is the remote nature of offshore installations. The offshore rigs can be located hundreds of miles away from land, conducting complex and dynamic operations in harsh environments.

Operation failures in such installations may result in major consequences for human operators, the environment and process equipment. Safe and efficient teleoperation is critical for such unmanned facilities, securing benefit and optimal productivity at remote locations.

Requirements for Software Development

A mobile robot for inspection and manipulation in offshore environments may only be acceptable if it can be used without

expert knowledge but rather easily and intuitively as a daily-used tool. This implies that:
- the robot can be controlled manually, semi-automatically and autonomously;
- new inspection and manipulation tasks can be programmed quickly and without the assistance of specialists;
- anyone working next to the robot can interact with it safely.

Therefore, the control software for a mobile offshore robot should provide functions to
- navigate on the platform without collisions both in remote-controlled mode and in automatic mode, for example move to a given target autonomously;
- easily program typical inspection, monitoring and/or manipulation tasks;
- execute pre-programmed inspection, monitoring and/or manipulation tasks automatically;
- enable supervision and control of the robot from a remote location;
- display sensor data such as camera images, the current gas concentration or other sensor information on the remote screen;
- alert the remote operators when abnormal sensor values are detected;
- review sensor recordings of past autonomous inspection tasks at the remote screen.

Requirements for Robotic Systems

Because it is installed at remote and isolated places, offshore oil and gas platforms pose a challenging environment for their human operators due to the unsheltered maritime environment, heavy weather and unfriendly, often explosive, toxic and corrosive atmosphere. In order to apply mobile robotics technologies in offshore environments, a number of challenges that do not exist in any other application area of mobile robots must be overcome [10]

- Complex environments: Offshore installations contain complex structures such as pipes, flanges, tanks, steel frames, stairways and many more. These structures can be very hard to detect by sensor systems applied typically in mobile robotics. The installed sensors on the mobile offshore robot must be able to distinguish relevant structures in the environment such as obstacles and free passages. The process equipment on offshore platforms is distributed among different levels, often including intermediate or mezzanine levels. only stairs or ladders are available to move from one level to the other. An appropriate means to move the robot from one level to another is required for a single robot to cover all levels.
- Floors: Offshore platforms on the contrary mainly consist of plain steel floor and gratings. There are many small holes, sharp edges, slopes and steps up to several centimeters in height.
- Explosive atmospheres.
- Corrosive environments: splash salty water, salty air and corrosive chemicals.
- Temperature: the ambient temperature on offshore platforms shows significant variations, depending on the platform's location. The mobile offshore robot must thus be operable in temperature ranges between $-30°C$ and $+50°C$.
- Humidity: The relative humidity up to 100% and condensing.
- Other conditions: possibly highly radiant heat from process equipment, heavy precipitation, storms, and direct sunlight.

CURRENT ROBOTICS TECHNOLOGIES

There are different kinds of robots in the oil & gas industry. The remotely operated underwater vehicles are discusses first. Since there are many underwater pipeline problems, underwater pipeline repair robotic systems are developed to repair the pipelines. Mobile robot platforms for topside oil & gas platform inspection and operations are investigated by academia and industry. Several systems are presented to understand their features and specifications.

Remotely Operated Underwater Vehicles (ROVs)

ROVs are unoccupied, highly maneuverable underwater robots operated by an operator aboard a surface vessel. They are linked to the ship by different cables that carry electrical signals back and forth between the operator and the ROV. Most are equipped with video cameras and lighting systems. Sometimes additional equipment is added to expand the vehicle's capabilities, including still cameras, manipulators or cutting arms, water samplers, and instruments that measure water clarity, light penetration, and temperature. ROVs were developed for industrial purposes, such as internal and external inspections of pipelines and the structural testing of offshore platforms; however, they are now used for many applications including scientific ones. They have proven extremely valuable in ocean exploration. They are also implemented for educational programs at aquaria.

ROVs vary in size from that of a bread box to a small truck. Deployment and recovery operations range from simply dropping the ROV from a small boat to complex deck operations using large winches for lifting and Aframes to take the ROV back onto the deck. In most cases, however, ROV operations are simpler and safer to conduct than many types of occupied-submersible or diving operations.

The disadvantages of using a ROV include: (a) the human presence is lost, making visual surveys and evaluations more difficult; (b) the lack of freedom from the surface due to the ROV's cabled connection to the ship. An ROV operator controls the vehicle from a system on board the ship using a joystick, a camera control, and a video monitor. The operator moves the vehicle and the camera to desired locations. The vehicle's depth, direction, and geographic position (latitude and longitude) are also recorded.

ROVs are often kept aboard vessels mounting submersible operations for several reasons. In the event that a submersible becomes entangled or otherwise incapacitated, an ROV can investigate the scene to help the operators make decisions. If appropriate, cutter blades can be deployed to the manipulator arm to free the sub. If a sub loses power and is not able to surface, the ROV's manipulator arm can grab onto the sub. The deck crew can then bring it to the surface.

ROVs also support exploration and science objectives. When the submersible cannot be used because of weather or maintenance problems, the ROV often can take its place. It can also be used to investigate questionable dive sites before a sub is deployed, limiting risk to the expensive subs and their pilots.

There are different kinds of ROV manipulators developed [11] -[13] . One example is the TITAN manipulator [11] Constructed from titanium, the TITAN is uniquely capable of withstanding the industry's harsh and repetitive needs. Some features are:

- Seven degrees of freedom;
- Durable and reliable in harsh subsea environments;
- Large operating envelope;
- High lift-to-weight ratio;
- Depth rating from 4000 m to 7000 m;
- Titanium construction;
- Hydraulic.

The master controller of the ROVs contains function keys for selecting configuration options and a display for viewing diagnostic and status information. The controller's advanced operational features include individual joint operation, position scaling (altering the ratio of master arm movement to manipulator arm movement), programmable routines, incremental gripper movement, individual joint diagnostics, and automatic error checking.

Deep Water Pipeline Repair Robotic Systems

Since there are many requests to repair deep water pipelines from the international oil and gas companies, StatOil [14] and Cheron [15] have developed pipeline repair robotic systems. The pipeline repair robots can work for water depths down to 1000 meters.

SINTEF Topside Robotic System

Because it is difficult to access subsea installations, normally-unmanned automated topside platforms may be an alternative through increased accessibility for large maintenance operations. Moreover, topside platforms may statistically recover up to 22 percent more of the oil or gas than a subsea alternative in a reservoir [7] . Due to these reasons, SINTEF worked with Norwegian energy producer StatoilHydro on a remotely operated offshore topside platform [8] [10] [16] in the robotic lab facility in Trondheim, Intelligent and reliable robotic and instrumentation systems has been developed to enable onshore operators to monitor and control all of the platform's processes.

Compared to their current manned counterparts, the normally-unmanned oil platforms offer potentially significantly lower commissioning and operation costs. Remote inspection and maintenance (I&M) operations can be performed on offshore oil and gas platforms as an alternative to traditional offshore platforms. The platforms separate the work area accessible by human operators from a closed permanently unmanned area (PUA) serviced by robots. Some important scheduled I & M operations inside the PUA can be performed by robots to replace human operators such as gauge readings, valve and lever operations and monitoring gas level, leakage, acoustic anomalies and surface conditions. Standard 6-axes robot manipulators (Kuka KR-16) are mounted on a mobile platform to automatically connect to custom-built tools and sensors such as vibration-measurement sensors, a valve-operating tool, and a lid operation tool. The operators will remain on land and are able to perform different tasks, reducing both risks and costs.

Fraunhofer Inspection Robot

The Fraunhofer inspection robot [17] was developed and tested by the Fraunhofer Institute of Manufacturing Engineering and Automation (IPA) for inspection. The robot uses specifically shaped objects such as pipes and poles as well as stripes of reflective tape applied to the environment for localization. It was tested continuously for 12 hours per day in hazardous locations and in tropical environment with 35+°C ambient temperature, up to 90% rel. humidity, and direct sunlight. A laser scanner is used to perceive its environment. A six-axis lightweight arm installed on the robot carries a camera to perform visual inspections. The robot platform has various application-specific sensors, such as a stereo microphone as well as gas and fire sensors. Wireless LAN and Bluetooth are deployed to enable the robot to communicate with the central control PC and with a mobile operator control device.

The robot is capable of safe navigation in offshore environments, which enables the robot to autonomously record sensor data at key locations or continuously monitor sensor data along a predetermined path. A manual operator control device, such as an explosion-proof PDA, can be used to teach the robot to perform inspection tasks. The robot autonomously executes the inspection tasks after the inspection tasks have been taught. The sensor data are recorded and displayed at the central control PC. The operators in the central control room are able to supervise all relevant sensor data in real time. In addition, the robot can be teleoprated to assist the user by analyzing data of its environment sensors. The user drives the robot around the oil & gas platform in order to get close to objects that need to be inspected or manipulated. A graphical user interface can display the objects detected by the robot through its sensors. The robot can then perform the planned tasks automatically such as executing movements or grasping operations to solve tasks, e.g. positioning its camera in front of a gauge or turning a hand wheel.

CMU Inspection Robot

The sensabot [18] was developed by Carnegie Mellon University, supported by Shell. Sensabot was designed for severe weather, pack ice, and temperatures that range from over 100°F (40°C) to below −31°F (−35°C). On top of that, the robot should work in an environment with a corrosive, toxic, foul-smelling gas and explosive methane gas. Sensabot meets the International Electrotechnical Commission (IECEx) standards for electronic equipment and ANSI safety standards for guided industrial vehicles. The safety systems on Sensabot reduce the risks of environmental hazards and operating around personnel.

The mobile base of the Sensabot can deal with concrete, gravel, and steel gratings and tackles ramps, light snow, and slippery surfaces. It can go through human-sized walkways. It can access to multiple levels of a facility via ramps, elevators, and a cog rail systems. There are different inspection sensors on Sensabot including (a) detectors to measure concentrations of gasses such as methane; (b) a powerful pan/tilt/zoom camera for remote operators to visually inspect pipes, valves and fittings for corrosion and other types of damage; (c) temperature sensors to help operators spot overheating equipment; (d) vibration sensors to enable operators to monitor the condition of pumps, motors, and bearings; (e) a microphone to help operators detect audible problems with machinery; (f) ladar and video to detect obstacles in its vicinity (such as equipment that lies in its path).

FURTHER RESEARCH OPPORTUNITIES

Because of the challenges and special requirements in the oil & gas industry, there are many exciting research opportunities for robotics and automation. This section discusses the research areas in robot manipulators, mobile platforms, teleoperation and automated equipment.

Manipulator

Currently, pipe handling robots and ROV arms are two types of manipulators used in oil and gas. As people are looking for more automated oil rigs, specialized robot arms should be developed. The robot arms should be able to tolerate the harsh environments in the oil and gas areas. Besides that

- The inspection robots should have a large workspace since the oil and gas equipment is typically much bigger than that in automotive manufacturing;
- The maintenance and/or repair robots should be able to handle big payload;
- The operation robots should have different tools;
- Light weight manipulator with big workspace for inspection;
- Heavy duty manipulator for operations.

Mobile Platform

ROVs are widely used in subsea operations to install equipment, perform maintenance and repair tasks etc. However there are many disadvantages for the current ROVs:

- Cables: the cables with the ROVs are easily tangled because they extend from the vessel to where the ROV is located. Since there are many objects in the sea, the cables could be tangled. Once ROV is tangled, the cables have to be cut.
- ROVs are not easy to operate. Even though there are some training systems to help the operators, it is a difficult task for the operators.

For the future research in the ROV, the opportunities could be:

- ROVs without using long cables from a vessel.
- Semi-autonomous ROVs. The ROVs can guide itself to avoid obstacles and locate the tasks.
- Autonomous ROVs. The ROVs are preprogrammed to perform given tasks before operations.

The ROVs with intelligence can make decisions and perform given tasks. Besides ROVs, mobile platforms will have many applications in oil and gas:
- Topside inspection and monitoring;
- Topside maintenance and repair;
- Topside operations.

The future research for the mobile platforms could be:
- Mobile platforms should be able to tolerate harsh environments;
- Mobile platforms should be able to anchor themselves automatically;
- Teleoperated mobile platforms could be more feasible in a short term.

Due to the development of unmanned platforms on the surface and teleoperated seafloor drilling rigs, mobile manipulators will parallel deep sea ROVs in maintaining platforms, where it is reasonable to assume the industry will see many more unmanned platform projects in the very near future.

Teleoperation

Teleoperation will still be the enabling technology in a short term because the oil & gas industry hesitates about using fully automated robotic system on the oil and gas operations. Different teleoperation technologies should be developed for different applications, for example, mobile platforms on the rig and subsea mobile platforms.

Subsea Robots

Half of the world's remaining oil reserves lie beneath the oceans' floors. Getting to them presents distinct engineering challenges, ranging from weather and ice-related issues to the environmental and human safety risks. Automated, robot-operated, unmanned drilling rig that sits on the bottom of the sea would resolve the problems related to weather-tossed seas, darkness, floating icebergs, and pack ice. Instead of building an offshore surface platform to drill for oil and gas at the seabed, up to 3000 meters below, exploration companies can use the Seabed Rig system to drill on the sea floor using robotic systems, with only a support vessel at the surface. The Seabed Rig also eliminates the potential for human error, is safer, and costs less than conventional drilling platforms.

There are many opportunities for robotic system in seabed operations. Besides drilling, other possibilities are:
- Automated seabed maintenance and repair system;
- Automated seabed inspection system;
- ROVs.

Automated Equipment

Because the oil and gas industry does not require highly flexible systems, hard automation could be better solutions for the industry. For example, seabed drilling system is more like a hard automation system, instead of robotic system. Furthermore, Robotic drilling systems, such as the one developed by Seabed Rig, would appear to be extremely useful systems for simplifying the drilling process and ensuring that its accuracy is to the utmost degree. Most drilling processes are extremely labor intensive and require continual human supervision, which could be easily remedied by robotic systems.

Related Research Topics

Since an offshore platform is quite complex, there are many wires used to transmit signals or transfer power. Data communication is a critical issue. Some potential research areas are:
- Sensor network;
- Wireless data communication;
- RFID applications;
- Data management.

Since the paper focuses on exploring robotic applications in the oil & gas industry, the above research areas will be discussed in our future research.

CONCLUSIONS

There is little doubt that the oil & gas companies would benefit greatly from the use of more intelligent technologies, not only increasing their efficiency, but also cutting down on human risk factors that are ever present in rig environments.

There is a clear incentive for oil and gas companies to automate their oil and gas facilities, starting with isolated operations, such as pipe handling and assembly for drilling and tasks related to rig operations. These examples represent high-risk operations for humans and therefore provide opportunities to improve health, safety and environmental (HSE) performance.

In addition to productivity and efficiency gains, robots used for high-risk tasks will also lead to improvements in HSE performance. Such tasks are not necessarily always predictable and represent unusual robot activities. The robot will therefore require features that extend the "eyes, ears, and hands" of the human decision makers as they carry out inspections and maintenance operations on the process infrastructure.

Reduced commissioning and operation costs, together with improved Environmental, Health and Safety (EHS) are some of the potential benefits of having normally unmanned topside oil platforms. However, such oil & gas platforms require advanced methods and tools for remote control and monitoring of inspection and maintenance operations.

In this paper, a brief introduction to the oil & gas industry processes is performed. The challenges and requirements for the robotics and automation equipment are explored. Future research opportunities including robot manipulator, mobile platform, teleoperation, and subsea robotics are discussed. Overall, we postulate that there are many opportunities in the oil and gas industry and some research is currently in progress to develop robotic and automation applications. Therefore, it is an optimal time to develop robotic and automation systems that can satisfy the oil & gas requirements.

ACKNOWLEDGEMENTS

The authors would like to thank ABB Corporate Research Center, ABB Inc. Windsor, Connecticut, for supporting the study.

REFERENCES

1. World Nuclear Association (2012) Uranium, Electricity and Climate Change.http://www.world-nuclear.org/info/Energy-and-Environment/Uranium, Electricity-and-Climate-Change/
2. Upstream (Petroleum Industry). (2014).http://en.wikipedia.org/wiki/Upstream_(petroleum_industry
3. Upstream Oil and Gas. (2014). http://www.upstreamoilandgas.com/
4. Gary, J.H. and Handwerk, G.E. (1984) Petroleum Refining Technology and Economics. 2nd Edition, Marcel Dekker, Inc., New York.
5. GE Energy. (2011). http://www.ge-energy.com/solutions/index.jsp#tabs-industries
6. US Environmental Protection Agency (2000) US EPA Office of Compliance Sector Notebook Project: Profile of the Oil and Gas Extraction Industry. US Environmental Protection Agency, Washington DC.
7. Meeting the Challenges of Today's Oil and Gas Exploration and Production Industry.http://www-935.ibm.com/services/us/gbs/bus/pdf/g510-3882-meeting-challenges-oil-gas-exploration.pdf

8. Transeth, A.A., Skotheim, O., Schumann-Olsen, H., Johansen, G., Thielemann, J. and Kyrkjebo, E. (2011) A Robotic Concept for Remote Maintenance Operations: A Robust 3D Object Detection and Pose Estimation Method and a Novel Robot Tool. The International Conference on Intelligent Robots and Systems, Taipei, 18-22 October 2010, 5099-5106.
9. The Robotized Field Operator. http://www.isa.org/InTechTemplate.cfm?template=/ContentManagement/ContentDisplay.cfm&ContentID=79568
10. Kyrkjebo, E., Liljebaäck, P. and Transeth, A.A. (2009) A Robotic Concept for Remote Inspection and Maintenance on Oil Platforms. Proceedings of ASME 28th International Conference on Ocean, Offshore and Arctic Engineering (OMAE 2009), Hawaii, 31 May-5 June 2009, 667-674.
11. Schilling Robotics. (2014).http://www.schilling.com/products/manipulators/Pages/default.aspx
12. Atlas Hybrid Manipulator. (2014). http://www.oceaneering.com/rovs/rov-technologies/atlas-hybrid-manipulator/
13. ECA Robotics. (2014). http://www.eca-robotics.com/en/control-command-security/robotics/2.htm
14. Statoil Developing Deepwater Pipeline Repair Robot. (2014). http://www.offshore-mag.com/articles/2007/03/statoil-developing-deepwater-pipeline-repair-robot.html
15. Chevron's Deepwater Pipeline Repair System Revealed. (2013). http://oil-geneva.blogspot.com/2013/02/chevrons-deepwater-pipeline-repair.html
16. Robots Taking over the Job on Offshore Drilling Platforms. (2014).http://www.sciencedaily.com/releases/2007/12/071221230852.htm
17. MIMROex. Fraunhofer Institute of Manufacturing Engineering and Automation.http://www.ipa.fraunhofer.de/MIMROex.2297.0.html?&L=2
18. SensaBot Inspection Robot. Robotics Institute, Carnegie Mellon University.http://www.ri.cmu.edu/research_project_detail.html?project_id=752&menu_id=261

Chapter 8

Vibration of Slender Structures Subjected to Axial Flow or Axially Towed in Quiescent Fluid

L. Wang and Q. Ni

Department of Mechanics, Huazhong University of Science and Technology, Wuhan 430074, China

ABSTRACT

The vibrations and stability of slender structures subjected to axial flow or axially towed in quiescent fluid are discussed in this paper. A selective review of the research undertaken on it is presented. It is endeavoured to show that slender structures subjected to axial flow or axially towed in quiescent fluid are capable of displaying rich dynamical behavior. The basic dynamics of straight and curved pipes conveying fluid (with or without motion constraints), carbon nanotubes conveying fluid, tubular beams subjected to both internal and external flows in axial direction, slender structures in axial flow or axially towed in quiescent fluid, cylindrical shells conveying or immersed in axial flow, solitary plate or parallel-plate assembly in axial flow; linear, nonlinear, and chaotic dynamics; these and many more are some of the aspects of the problem considered.

INTRODUCTION

The study of flow-induced vibrations of slender structures has been intensified in the past decades. This may be partly due to the increased need for stability and reliability, especially in the power generating industry where repeated equipment failures have evidenced the inadequate state-of-the-art. Thus, it has now become increasingly important to try to understand and to be able to predict the dynamical behaviour of slender structures in flow, such as what might be found in mechanical equipments, nuclear reactors, heat exchangers, steam generators, ocean mining pipes and drill-strings [1–3]. Unlike the case of vibrations of slender structures induced by cross-flow, the study of dynamical behaviours of such structures induced by axial flow is a relatively new phenomenon, beginning seriously in the 1960s [4, 5]. Although most failures are associated with the cases of cross-flow, the conditions of axial flow have also been shown to be of importance. Moreover, quite apart from practical considerations, these problems are of sufficient intrinsic interest, in the realm of dynamics of various dynamical systems subjected to gyroscopic forces (e.g., axially accelerating belts), to merit study for their own sake.

Motivated by the quest for a fundamental understanding of fluid-structure interactions as well as by applications in several areas of engineering, the vibrations of axially moving slender beams or cylinders in fluid has also attracted the attention of several investigators [6–8]. As can be expected, the dynamics of slender structures axially towed in fluid should be different from that of slender structures conveying fluid or immersed in axial flows.

The current paper attempts to present a selective review of the published literature, emphasizing the work dealing with slender structures subjected to axial flow or axially towed in fluid, specifically the vibrations and stabilities of (i) pipes conveying fluid, both straight and curved (ii) carbon nanotubes conveying fluid (iii) tubular beams subjected both internal and external axial flows (iv) cylindrical shells conveying fluid (v) plates in axial flow and (vi) slender structures axially towed in quiescent fluid. The main purpose of this article is to review the recent literature which is relevant to all aspects and ramifications of slender structures subjected to axial flow or axially towed in fluid. In order to display some new results reported on this problem, the related background theory and some of the early work will also be discussed.

DYNAMICS OF STRAIGHT PIPES CONVEYING FLUID

Since the dynamics of straight and curved pipes, cantilevered and supported pipes, are fundamentally different, they will be treated separately. In each case the linear dynamics will be treated first.

Linear Dynamics of Straight Pipes Conveying Fluid

The Simplest Equation of Motion

Consider a straight pipe conveying fluid (Figure 1). If externally imposed tension, internal damping, gravity, a possible elastic foundation and pressurization effects are either neglected or absent, the linear equation of motion can be written in a particularly simple form [9]

$$EI\frac{\partial^4 w}{\partial x^4} + MU^2\frac{\partial^2 w}{\partial x^2} + 2MU\frac{\partial^2 w}{\partial x \partial t} + (M+m)\frac{\partial^2 w}{\partial t^2} = 0, \tag{1}$$

where EI is the effective flexural rigidity of the pipe, m is the mass of the pipe per unit length, M is the mass of fluid per unit length, flowing with a steady flow velocity U, and w is the lateral deflection of the pipe; x and t are the axial coordinate and time, respectively. The fluid forces are modeled in terms of a plug flow model, which is the simplest possible form of the equation of motion for straight pipes conveying fluid. A more detailed treatment of the linear equation of motion was given in [9]

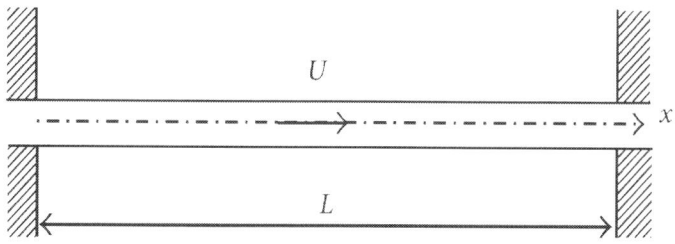

Figure 1: Schematic of a fluid-conveying straight pipe with both ends supported.

The various terms in (1) may be defined, sequentially, as the flexural restoring force, a centrifugal force, a Coriolis force, and the inertia force.

Straight Pipes with Supported Ends

The linear dynamics of the system for a pipe simply supported at both ends is very clear now. After a Galerkin discretization, the equation of motion, as given by (1), can be solved by considering a characteristic eigenvalue problem. The eigenfrequencies (denoted by ω in this paper) of the pipe system are generally complex. For the sake of convenience, we may define U as the dimensionless flow velocity, that is, $\bar{U} = (M/EI)^{1/2} LU$. Note that Re (ω) is the oscillation frequency, while Im(ω) is related to the damping, the damping ratio being Im(ω)/Re(ω). It has been found that divergence in the first mode occurs at $\bar{U}_c = \pi$ and in the second at $\bar{U}_c = 2\pi$. Linear theory predicts a coupled-mode flutter for $\bar{U} > 6.375$.

In [9], a similar result was obtained for a clampedclamped pipe. In this case, the first-mode divergence occurs at $\bar{U}_c = 2\pi$; then the system is restabilized at $\bar{U} \approx 9$. Again, still according to linear theory, another form of postdivergence (coupled-mode) flutter was predicted at $\bar{U} \approx 9.3$.

However, the physical existence of this postdivergence flutter instability is questionable, since the linear equation of motion cannot be used to provide reliable information once the displacements become large. Therefore, this postdivergence flutter would have to be verified by nonlinear theory, as will be discussed later.

Straight Pipes with Clamped-Free Ends

Unlike the straight pipes with both ends supported, the linear dynamics for a cantilevered system may show significant difference [10]. Up to now, it has been reported that the effect of increasing $\bar{U}$ provided it remains small, is to generate flowinduced damping in the system. For a relatively small flow velocity, it was shown that the value of Im(ω) remains positive. For increasing values of $\bar{U}$ (e.g., $\bar{U} \approx 4$), however, this damping begins to be attenuated. The damping eventually vanishes (at $U \approx 5.6$) and then becomes negative. This implies that a single-degree-of-freedom (DOF) flutter via a Hopf bifurcation occurs. The presence of nonzero damping at $\bar{U} = 5.6$, therefore, merely postpones the onset of flutter

In the last two decades, a vertically cantilevered pipe conveying fluid upwards was also studied by several investigators [11–14]. It ought to be recalled that this problem is not wholly academic. One important application may be in the ocean mining industry. The ocean mining of manganese nodules, by essentially vacuuming the sea floor from a surface vessel, thus involves a flexible, long pipe. In this case, the vertical pipe is always aspirating fluid; thus, the dynamics may be different from that of the system conveying fluid downwards, as discussed in the foregoing.

Up to now, in several experiments, it was observed that a fluid-aspirating pipe is stable for small flow velocity. However, theoretical results [14] have shown that flutter might occur even for pipes aspirating fluid with sufficiently low-flow velocity. In 2005, a reappraisal of why aspirating pipes do not flutter at infinitesimal flow is made by Païdoussis et al. [14]. In that paper, the linear equation of motion of a cantilevered pipe aspirating fluid was written as

$$EI\left(1 + \alpha^* \frac{\partial}{\partial t}\right) \frac{\partial^4 w}{\partial x^4} + \left[MU^2 - (\bar{T} - \bar{p}A)\right] \frac{\partial^2 w}{\partial x^2}$$
$$+ 2MU \frac{\partial^2 w}{\partial x \partial t} + c \frac{\partial w}{\partial t} + (M + m + M_a) \frac{\partial^2 w}{\partial t^2} = 0 \quad (2)$$

In which the added mass per unit length M_a of the ambient fluid has been included; c is the viscous damping coefficient, A is the cross-section-area of the fluid, α∗ is the coefficient of Kelvin-Voigt internal dissipation, $\bar{p}$ is the global pressurization, and T the externally imposed tension.

From (2), if the dissipation is absent or neglected, it can be easily verified that instability is possible for small internal flow velocity. If, however, a realistic amount of dissipation is taken into account, the straight pipe was found to remain stable up to flow velocities covering the range of practical interest (for the ocean mining application, e.g.). Thus, it has now become imperative to try to understand the stability mechanism of pipes aspirating fluid. However, this question remains unresolved. In [14], it has been suggested that more precise assumptions made on the intake flow structure should be developed in the near future. Was the observed stability only due to dissipation? Or was it because the flow at the intake is such as to make flutter impossible? To help get to the bottom of things, a CFD study of the flow field would be helpful, as mentioned in [14].

Nonlinear Dynamics of Straight Pipes Conveying Fluid

Straight Pipes with Supported Ends

Presuming the existence of periodic motions, the rate of work done by the fluid on the pipe over a period of oscillation T may be obtained from (1) [13]

$$\Delta W = -MU \int_0^T \left[\left(\frac{\partial w}{\partial t}\right)^2 + U\left(\frac{\partial w}{\partial t}\right)\left(\frac{\partial w}{\partial x}\right) \right] \Big|_0^L dt. \tag{3}$$

Clearly, if both ends of the straight pipe are positively supported, then $(\partial w/\partial t) = $ at both ends.

$$\Delta W = 0. \tag{4}$$

This implies that self-excited oscillatory motion (flutter) is not possible for pipes with both ends supported. However, this does not mean that the system remains stable even for sufficiently high U. The $+MU^2\,(\partial^2 w/\partial x^2)$ term can be viewed as an effective compression associated with the exiting fluid momentum at the downstream end. This might be linked to a slender column subjected to a compressive force. For high enough U, therefore, the system would lose stability by buckling (static divergence).

In fact, even in the context of linear theory, the existence of flutter is problematic for a pipe with positively supported ends. This delicate question was first addressed by Done and Simpson [15]. The question of the existence of postdivergence flutter (coupled-flutter) in this system has been answered by Holmes and his coworker [16–18]. Reference [17] is categorically entitled "Pipes supported at both ends cannot flutter"—for pipes positively supported at both ends, that is, where axial sliding is not permitted. Obviously, this important conclusion was based on analyzing the nonlinear dynamics of fluidelastic systems.

All the work mentioned in the last paragraph relates to theoretical work. In fact, the postdivergence flutter has never been observed experimentally in pipes with both ends supported, though the loss of stability by divergence is easily observable. Perhaps this is the most potent evidence of nonoccurrence of postdivergence flutter in pipes with both ends supported.

Moreover, parametric resonances may occur if the flow in the pipe is not wholly steady but contains periodic pulsations. Recently, quasiperiodic and chaotic motions were detected in pipes conveying fluid with both ends supported; see, for example, [19, and 20].

Straight Pipes with Clamped-Free Ends

For a cantilevered pipe system, it is assumed that the free-end is at x = L. Then, one obtains [13]

$$\Delta W = -MU \int_0^T \left[\left(\frac{\partial w}{\partial t}\right)^2 + U\left(\frac{\partial w}{\partial t}\right)\left(\frac{\partial w}{\partial x}\right) \right] \Big|_0^L dt \neq 0.$$

(5)

By analyzing the above equation, it is clear that $\Delta W < 0$ for $U > 0$ and small, and free motions of the pipe are damped. If, however, U is sufficiently high, while over most of the cycle $(\partial w/\partial x)_L$ and $(\partial w/\partial t)_L$ have opposite signs, then one can obtain $\Delta W > 0$. This implies that the pipe will gain energy from the flowing fluid, and hence free motions will be amplified The nonlinear dynamics of straight pipes with clamped-free ends is concordant. If we denote the onset of flutter as a Hopf bifurcation, the Hopf bifurcation gives rise to limit cycle motions, which is also what is observed experimentally. In this case, however, the dynamics can be much more complex. The Hopf bifurcation can be either supercritical or subcritical, depending on the parameter of mass ratio $\beta(= M/(M + m))$ and a parameter involving the friction coefficient and the slenderness of the pipe [21]. Moreover, the limit cycle motion can be either two or three dimensional [22], again depending on β.

The chaotic dynamics of cantilevered systems was also investigated extensively in the past two decades (see, e.g., [23–32]). In the first such study, Tang and Dowell [23] considered a cantilevered pipe with an inset steel strip and two equispaced permanent magnets on either side of the free-end, thus exerting nonlinear forces on the strip and buckling it into one of the two potential wells on either side of the pipe. If the flow velocity is increased sufficiently above the critical value for flutter about the buckled state, numerical results will show that the cantilevered system might display chaotic motions. This autonomous system was studied only briefly and theoretically. A more extensive theoretical and experimental study was made of a pipe system with external excitation. In the experiments, the pipe was excited by a shaker. Once again, chaotic motions were detected when the amplitude of the external excitation was sufficiently higher than the threshold value of this force for chaos. The chaotic motions, strongly influenced by the flow velocity, have also been observed experimentally.

At about the same time, Païdoussis and Moon [24] undertook a combined theoretical and experimental study of the autonomous system of a cantilevered pipe conveying fluid. In this case, however, the pipe is interacting with motion constraints somewhere along the length of the pipe. The only nonlinearity considered in the system was due to the nonlinear constraints, modeled by cubic springs. The experiments for a cantilevered pipe, with either air or water internal flow, showed that, when the flow velocity was sufficiently beyond the onset of flutter for the pipe, the pipe would impact on the motion constraints, thus introducing nonlinear force. The system became chaotic through the route of period doubling bifurcations. The analytical model, after Galerkin discretization to two DOFs., exhibited a similar behaviour. The same analytical model was further studied and some new results were obtained by Païdoussis et al. [25]. It was shown that, after the Hopf bifurcation, a symmetry-breaking transcritical-like pitchfork bifurcation occurs, followed by a sequence of period-doubling bifurcations, leading to chaotic motions. Sample results of phase portraits based on the equation of motion given in [25] are shown in Figure 2.

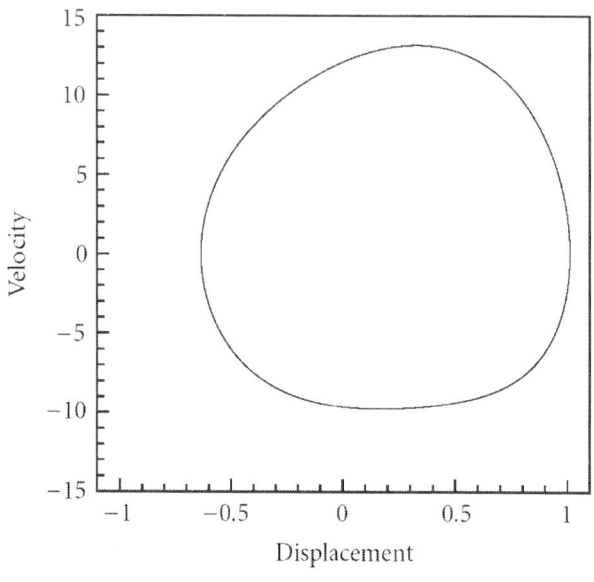

(a)

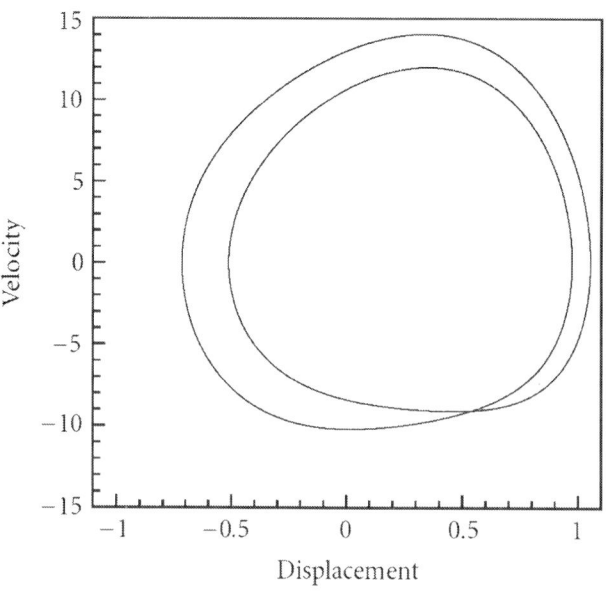

(b)

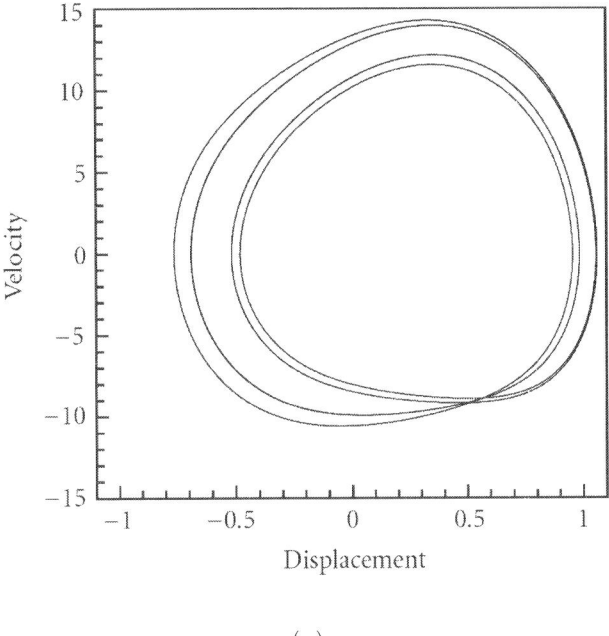

(c)

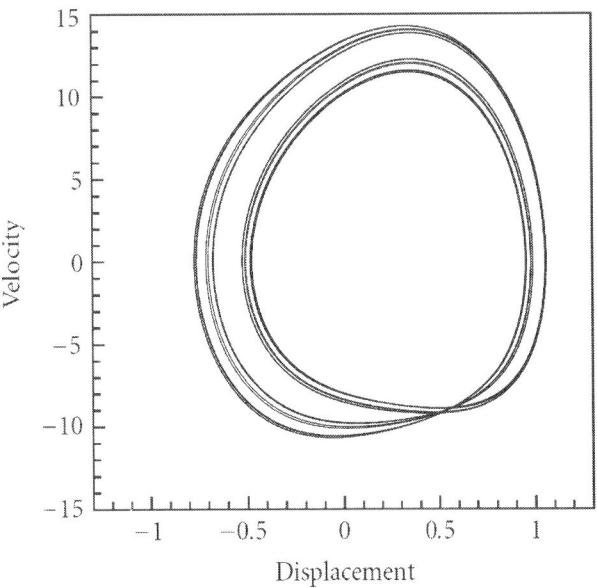

(d)

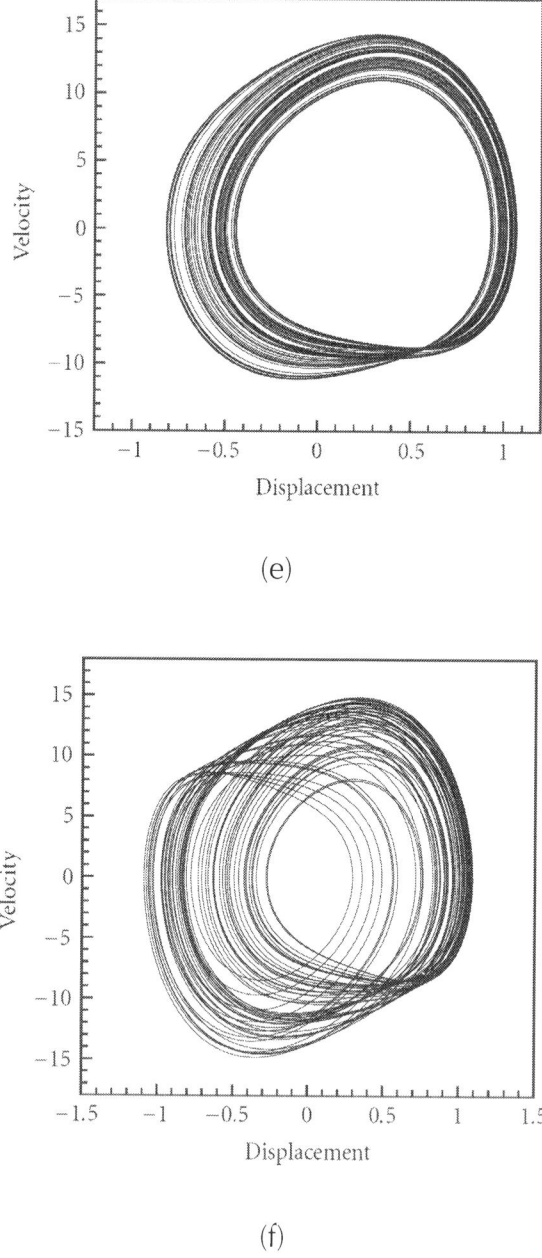

Figure 2: Theoretical phase portraits of the free-end of the pipe, with motion constraints modeled by a cubic spring and 2 d o f, for different flow velocities.

This same system was restudied with higher dimensional models (up to 7 d.o.f.) by Païdoussis et al. [26]. More significantly, the impact models for the motion constraints were improved. In this case, for the exact experimental parameters in the analytical model, excellent quantitative agreement (within 5–10%) was obtained.

Finally, by using the full nonlinear equations of motion [27], the same cantilevered system was re-examined, completing the circle of studies on this system. The post-Hopf dynamics predicted with lower d.o.f. (e.g., 3 d.o.f.) was quite different from that predicted with higher d.o.f (e.g., 4 d.o.f.). It was found that, the degree of agreement with experiment becomes excellent for at least 4 d.o.f.

Also, if the cantilevered system is standing, interesting dynamics (e.g., the pipe regains stability after flutter instability as the flow velocity is increased) may arise [28, 29]. However, in the study of [29], the only nonlinearity considered is the nonlinear force induced by the motion constraints.

Moreover, chaotic motions may also occur for a cantilevered system with an intermediate linear spring support, or with a mass added at the free-end [30–32]. In [31, 32], the pipe is allowed to vibrate in a 3D space. The intermediate spring supports were disposed in symmetrical fashion with respect to two axes. For the pipe with a four-spring array placed somewhere along the length of the pipe (but not close to the free-end), it was found that the system loses stability via a Hopf bifurcation (i.e., similarly to the planar motion of a cantilevered pipe). Again, a pitchfork, or symmetry breaking bifurcation was detected at a higher flow velocity. However, the pipe was predicted to oscillate asymmetrically; and symmetry is regained for a higher flow velocity. Interestingly, quasiperiodic oscillations, followed by chaotic oscillations, have been found in such a dynamical system with sufficiently high-flow velocity. However, if the array of four springs is positioned closer to the free-end and the total stiffness of the springs is much larger, the initial instability was predicted to be a pitchfork bifurcation.

DYNAMICS OF CURVED PIPES CONVEYING FLUID

One might have thought that a similar analysis of straight pipes conveying fluid can be extended to curved pipes. This is not so, however. Actually, in contrast to the systems reported so far, until 1988, there remained considerable confusion and uncertainty as to the vibrations and stability of curved pipes conveying fluid.

Work on the vibration and stability of curved pipes conveying fluid appears to have started in the 1960s. Some of the key contributions in this realm were made by, but not limited to, Svetlitskii [33–35], Chen [36–38], Doll and Mote Jr. [39, 40], Hill and Davis [41], Dupuis and Rousselet [42], Misra et al. [43–45], Ni et al. [46], Qiao et al. [47, 48], and Jung and Chung [49]. The systems studied range from curved pipes shaped as circular arcs, L- or S-shaped configurations, analyzed by finite-element techniques [39–41, 43–45], transfer-matrix technique [42, 50] or differential quadrature method (DQM) [46–48, 51, 52].

Unlike the straight pipes, motions of curved pipes generally require four displacement variables (three displacements of the centerline and a twist angle) and hence at least four equations to govern the motions. For the circular-centerline pipes, depending on the initial shape and state of the system as well as assumptions made, it is often possible to decouple the motions into in-plane and out-of-plane motions.

There are three main theories available for predicting the stability and vibrations of curved pipes conveying fluid: the so-called "inextensible theories" [36, 37, 43, 46–48], the modified inextensible theory [44], and the complete "extensible" theories [33, 34, 40–42, 44, 45, 49]. The inextensible theories assume that the circular centerline of the pipe is essentially inextensible and all steady-state stress resultants are absent or neglected. The complete extensible theories, however, do not make this assumption and generally take into account the changes in form with increasing flow velocity, as well as the forces generated thereby; thus, the steady-state axial tension-pressure force has been considered. The modified inextensible theories take into account the initial stresses due to flowing fluid in a curved pipe.

The inextensible theories predict that curved pipes with both ends supported are subject to divergence at sufficiently high-flow velocities, similar to straight pipes. However, instability was predicted to be impossible by using the modified inextensible and the extensible theories. One interesting practical result is that the modified inextensible theory gives results very close to those of the complete extensible theory. For the cantilevered system, however, both the inextensible and extensible theories predict that the form of flutter instability might occur with sufficiently high-flow velocity. The reason may be that the steady-state axial tension-pressure force has a less pronounced effect on the dynamics of cantilever system [48].

Here, it should be pointed out that, the literature on the nonlinear dynamics of curved pipes conveying fluid is very limited. Dupuis and Rousselet [53] have undertaken a careful derivation of the nonlinear equations of motion by the Newtonian approach; however, their equations are not easy to solve because of their complexity. In 2005, by using the inextensible theory, Ni et al. [46] developed a cantilever model, in which a curved pipe is embedded in nonlinear foundations. Based on DQM discretization, the in-plane vibrations of the system were discussed, showing that chaotic transients could occur. Recently, Qiao et al. [47] investigated another cantilevered curved pipe conveying fluid with motion constraints (Figure 3) and explored some interesting dynamics. Again, the inextensible theory was utilized for the cantilevered system. As can be expected, the curved pipe would impact on the motion constraints when the deflection of the pipe becomes reasonably large due to increasing fluid velocity. The analytical model, after DQM discretization, exhibited various behaviors (see Figure 4). The route to chaos was shown to be via period-double bifurcations. This curved pipe model was further analyzed by Lin et al. [48] by applying an external excitation at the free-end of the curved pipe. In the forced pipe system, the routes to chaos were found to be via either period-double bifurcations or quasiperiodic motions. Therefore, the forced system can also display rich dynamics.

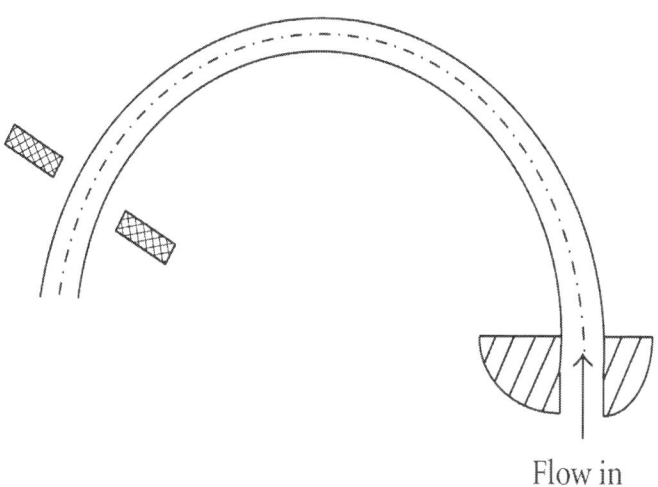

Figure 3: Schematic of a fluid-conveying curved pipe with motion constraints.

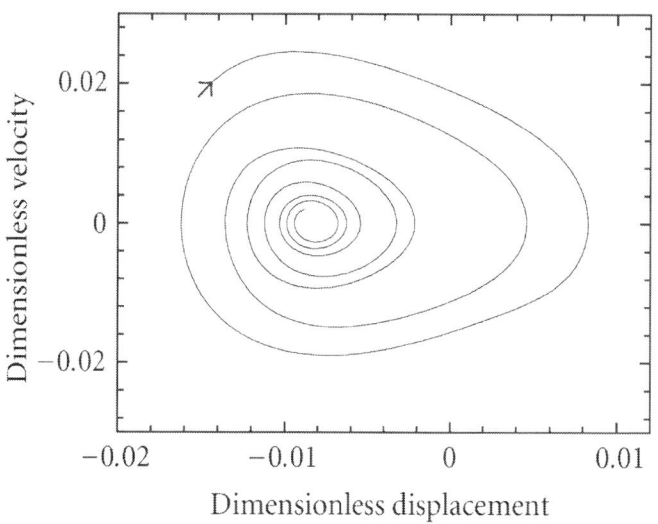

(a)

Vibration of Slender Structures Subjected to Axial Flow or ... 235

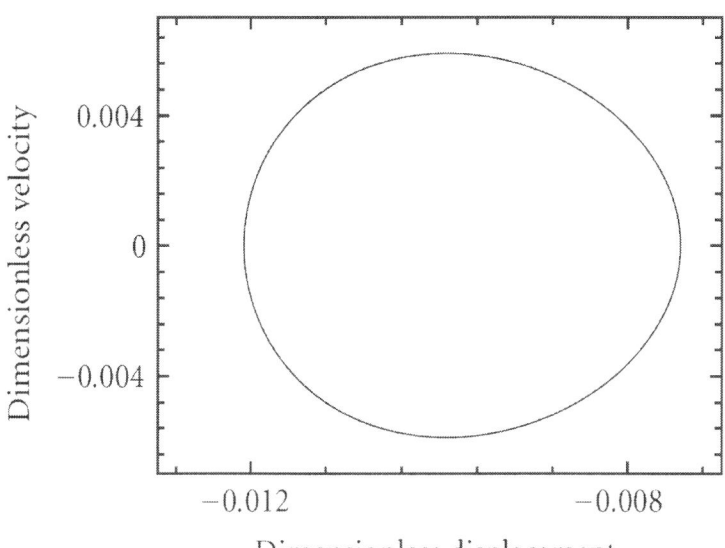

(b)

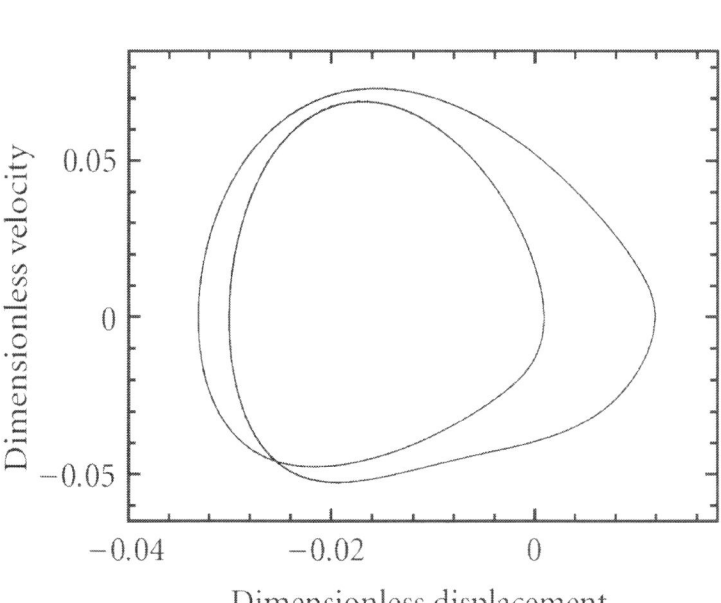

(c)

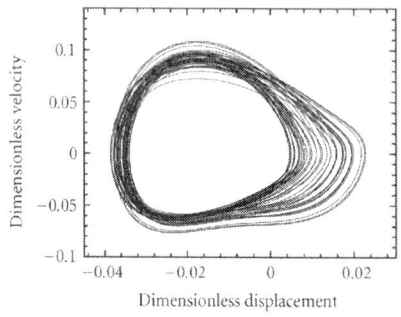

(d)

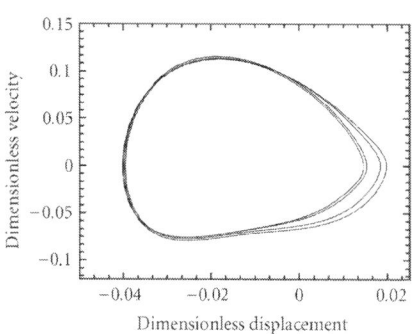

(e)

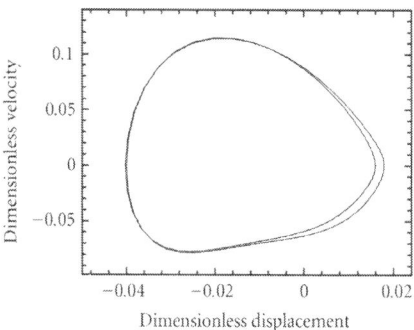

(f)

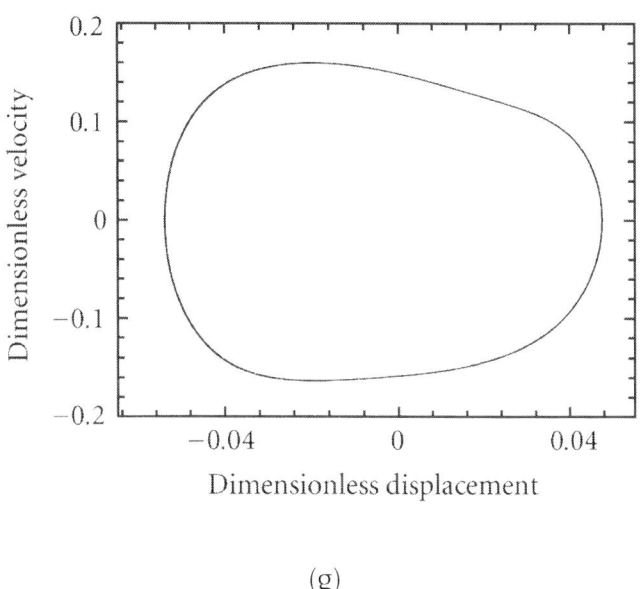

(g)

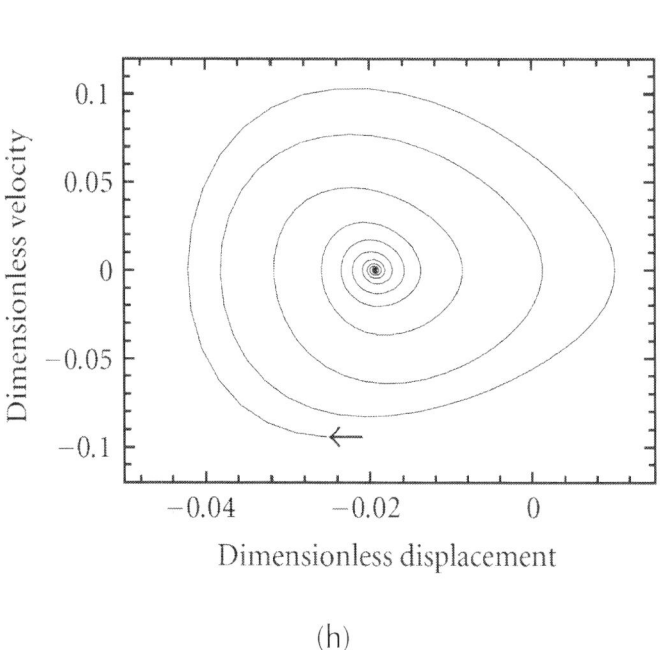

(h)

Figure 4: Theoretical phase portraits of the free-end of the curved pipe, with motion constraints modeled by a cubic spring, for different flow velocities; from [47].

Before closing this section, it ought to be noted that the fluid velocity flowing in curved pipes was assumed to be steady, in all the related work cited in the foregoing. More significantly, the geometric nonlinearities induced by the deformation of curved pipes have been neglected in [46–48]. If, however, the fluid velocity is not steady and the geometric nonlinearities are considered, the nonlinear dynamics of the curved pipe may be much richer.

VIBRATIONS OF NANOTUBES CONVEYING FLUID

After the invention of carbon nanotubes (CNTs) by Iijima [54], it has been shown that CNTs have good electrical and mechanical properties and so they have potential applications in design for nanoelectronics, nanodevices, nanocomposites, and so forth, [55]. Because of perfect hollow cylindrical geometry and high mechanical strength, CNTs hold substantial promise as nanocontainers for gas storage and as nanopipes for conveying fluid (such as gas or water) [56–58]. It is not surprising; therefore, fluid flowing inside CNTs has become an attractive research topic [59–62].

In an attempt to understand and be able to predict the fluid-structure interactions in CNTs conveying fluid, Tuzun et al. [63] developed molecular dynamics simulations of fluids flowing through CNTs. It was found that in a fluid conveying CNT system, the motion of the CNTs plays a significant role in the fluid flow. For example, a fluid flowing through the CNTs tends to straighten out the CNT as it flexes, and simultaneously excites longitudinal vibration modes of the CNTs.

Since molecular dynamic simulations are difficult for large-scale systems, continuum mechanics models have been utilized to investigate the vibrational behavior of CNTs conveying fluid [64–70]. Natsuki et al. [64] studied the wave propagation in single- or double-walled carbon nanotubes (DWCNTs) filled with internal flowing fluids by using an elastic shell model.

On the other hand, the transverse vibrations of fluid-conveying CNTs by using Euler beam theory have been studied recently. Yoon et al. [65, 66] have developed a single-elastic Euler beam model for vibrating CNTs containing flowing fluids, both for the cantilevered and supported systems. It was found that the effect of fluid flow velocity on the resonant frequencies of CNTs is significant. Structural instability of the CNTs could occur at a critical flow velocity. The critical flow velocity could cover the range of practical interest. However, the effects of flow velocity on the resonant frequencies and the instability of CNTs would be mitigated when a CNT is embedded in a surrounding elastic medium (such as polymer matrix). As pointed out by Yoon et al. [65], the available data in the literature showed that the flow velocity inside CNTs might range from 400 m/s to 2000 m/s, or even up to 50000 m/s, in spite of the fact that the available data for flow velocity of water inside CNTs (of very small innermost diameter) are much lower than this value. In 2007, Reddy et al. [67] investigated the effect of fluid flow on the free vibration and instability of fluid-conveying single-walled carbon nanotubes (SWCNTs), using both atomistic and Euler beam models. Wang et al. [68] reported some results of an investigation into the influence of internal flowing fluid on the coupling vibration of fluid-filled CNTs; again, the Euler beam model was used. Recently, Wang and Ni [69] further analyzed the fluid-conveying CNT model developed by Yoon et al. [65] and explored the possible postdivergence flutter existing in the same dynamical model. Of course, the postdivergence flutter was predicted by the linear theory. More recently, Wang et al. [70] considered the thermal effect on the vibration and instability of a fluid-conveying CNT. Based on the Euler beam theory, it was concluded that at low or room temperature the critical fluid velocity for nanotube including the thermal effect is larger than that excluding the thermal effect and increases with the increase of temperature change. At high temperature the critical fluid velocity for the nanotube including the thermal effect is smaller than that without considering the thermal effect and decreases with the increase of temperature change.

It is noted that [65–70] only considered the vibrations of single-walled carbon nanotubes conveying fluid. Continuing this line of investigation, the flow-induced vibrations of DWCNTs (Figure 5) or MWCNTs conveying fluid were also examined by using Euler beam theory [71–74]. Based on the multiple-elastic beam model, the van der Waals interaction between tubes has been accounted for, showing some important results.

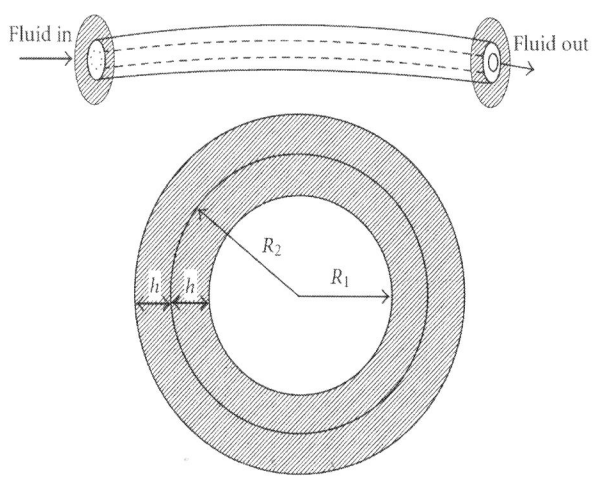

Figure 5: Schematic of a DWCNTs with internal fluid flow.

Finally, it ought to be stressed that, in [65–74], the vibration and stability of nanotubes conveying fluid were described by the Euler beam model, which also has been utilized to predict the dynamics of straight pipe conveying fluid, as discussed in Section 2. Therefore, the effect resulting from the small (nano-) scale on the vibrational properties of nanotubes conveying fluid has not been included so far. Although the Euler beam model and several other classical continuum models are relevant to some extent, the length scales associated with nanotechnology are often sufficiently small to call the applicability of continuum models into question. The main reason is that at small length scales the material microstructure (such as lattice spacing between individual atoms) becomes increasingly important. The effects of these small length scales can no longer be ignored. This has raised a major challenge to the classical continuum mechanics. Therefore, a possible solution is to develop some new fundamental theories based on the classical continuum models. Such new theories would account for the small length scales by incorporating information regarding the behavior of material microstructure.

DYNAMICS OF TUBULAR BEAMS SUBJECTED TO BOTH INTERNAL AND EXTERNAL AXIAL FLOWS

Because of important applications and academic requirements, the vibration and stability of a tubular beam system, subjected to both internal and external axial flows, have been studied by many investigators in the past decades. In Figures 6(a) and 6(b), two typical tubular beam systems are given. The configuration of Figure 6(b) thus resembles that of a drill-string with a floating fluid-powered drill-bit; for example, several related models developed in [3, 75–77].

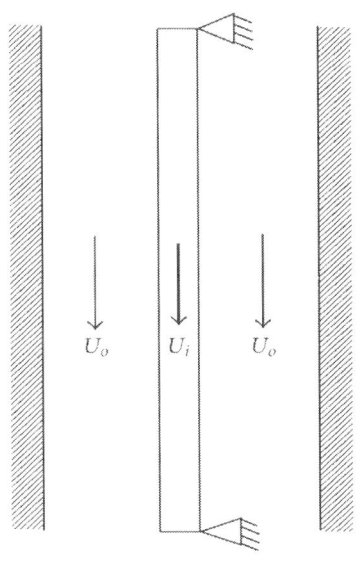

(a)

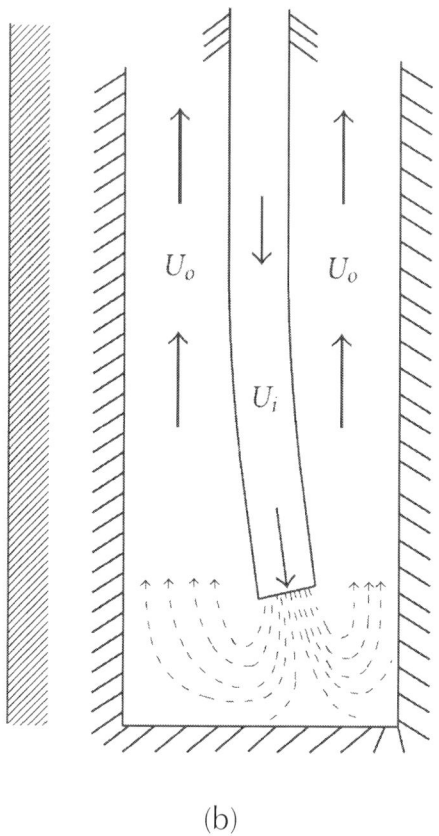

(b)

Figure 6: (a) Schematic of a tubular beam subjected to both internal and external axial flows; (b) the system considered in [83].

It is noted that, the tubular beams shown in Figure 6 are subjected concurrently to internal and external axial flows. In fact, the problem of a tubular beam subjected to both internal and external flows has been studied before, by many investigators. Cesari and Curioni [78] have studied the buckling instability in tubular beams subject to internal and external axial flows. Hannoyer and Païdoussis [79] combined theory and experiments on the linear vibrations and stability of a tubular beam with two supported ends or cantilevered, subject to both internal and external axial flows; they found multiple divergence and flutter instabilities. Theory and experiments were in quite good agreement. At about the same time,

Grigoriew [80] considered a drill beam with an initial curvature in the axial stream and analyzed its stability. Another notable work by Païdoussis and Besancon [81] discussed various aspects of the vibrations and stability of clusters of tubular beams conveying internal fluid and surrounded by a confined external axial flow. By calculating the eigenfrequencies of the tubular beam system and studying their evolution with various flow velocities of either internal or external fluid, the free vibrations were investigated. Wang and Bloom [82] formulated a mathematical model to study the dynamics of a submerged and inclined concentric tubular beam system with internal and external flows, the resonant frequencies of that system obtained and analyzed.

Recently, Païdoussis et al. [83] reported some interesting results on this problem. The basis of that work is Luu's thesis [84]. Luu [84] and Païdoussis et al. [83] differed from the work of [78–82] in two significant ways: first, the external and axial flows are countercurrent, and second the two flows are not independent of each other (see Figure 6(b)). In the study by Païdoussis et al. [83], a theoretical model was developed for the dynamics of a hanging tubular cantilever, centrally located in a cylindrical container, with fluid flowing downwards inside the cantilever. The internal fluid, after exiting from the free-end, is deflected at the bottom of the container, and thereafter flows upwards in the annular space between the cantilever and container. It is noted that the configuration developed in Païdoussis et al. [83] was inspired by the geometry of a drill-string with a drill-bit at the lower end.

The drill-string-like system considered in [83] consists of a uniform tubular beam of length L, external cross-sectional area A_o, flexural rigidity EI, and mass per unit length Mt, conveying downwards incompressible internal fluid of mass per unit length Mf, flowing axially with constant velocity U_i. The internal fluid leaving the lower end of the tubular beam then flows upwards with velocity U_o within an outer rigid channel. The linear equation of motion for this system can be written as [83]

$$EI\frac{\partial^4 w}{\partial x^4} + M_t\frac{\partial^2 w}{\partial t^2} + M_f\left(\frac{\partial^2 w}{\partial t^2} + 2U_i\frac{\partial^2 w}{\partial x \partial t} + U_i^2\frac{\partial^2 w}{\partial x^2}\right)$$

$$+ \chi\rho_f A_o\left(\frac{\partial^2 w}{\partial t^2} - 2U_o\frac{\partial^2 w}{\partial x \partial t} + U_o^2\frac{\partial^2 w}{\partial x^2}\right)$$

$$-\left[(T - A_f p_i + A_o p_o)_L + (M_t + M_f - \rho_f A_o)g(L-x)\right.$$

$$\left. -\frac{1}{2}C_f\rho_f D_o U_o^2\left(1+\frac{D_o}{D_h}\right)(L-x)\right]\frac{\partial^2 w}{\partial x^2}$$

$$+\left[(M_t + M_f - \rho_f A_o)g - \frac{1}{2}C_f\rho_f D_o U_o^2\left(1+\frac{D_o}{D_h}\right)\right]\frac{\partial w}{\partial x}$$

$$+ \frac{1}{2}C_f\rho_f D_o U_o\frac{\partial w}{\partial t} + k\frac{\partial w}{\partial t} = 0, \tag{6}$$

where $w(x,t)$ is the lateral deflection of the tubular beam and t is time; ρ_f is the mass density of the fluid; g is the acceleration due to gravity; C_f and k are the viscous damping coefficients; T_L is the axial tension, induced by the fluid pressure at the lower end; D_o is the outer diameter of the tubular beam and D_h is the hydraulic diameter of the annular channel flow. The definitions of p_{oL}, p_{iL} and χ can be found in [83]. Based on numerical calculations, Païdoussis et al. [83] have analyzed the evolution of the eigen frequencies. It was found that, if the annular space is wide, the dynamics is dominated by the inside flow (i.e., the flow within the drill-string), for low-flow velocities, the flow increases the damping associated with the presence of the annular fluid; if the annular space is narrow, however, the annular flow is dominant, tending to destabilize the system, giving rise to flutter at remarkably low-flow velocities.

From the viewpoint of string-drill dynamics these results are interesting for the following reason: it was shown that, even if the drill-bit never makes mechanical contact with the drill-string, the system experiences flutter type of instability; and hence the string would soon touch the surrounding walls. In a real system, however, effective contact between the drill-bit and the drill-string is inevitable, and so the dynamics is more likely to resemble those of a pipe with clamped or pinned ends [83].

DYNAMICS OF CYLINDRICAL SHELLS SUBJECTED TO AXIAL FLOW

Of academic and practical interest is the dynamics of thin-wall pipes conveying fluid. For very thin pipes conveying fluid, Païdoussis and Denise [85, 86] accidentally found that this type of pipe systems are not only subject to beam-type, but also to shell-type instabilities. For sufficiently high-flow velocities, if the thin pipe is relatively short, then the instability observed is not one of lateral motions of the pipe (n=1, where n is the circumferential mode number), but rather involves deformation of the pipe cross-section; that is, the instability is associated with a shell-type breathing mode (typically n = 2 or 3). When the diameter of the middle surface of the thin-wall pipe is relatively large, the problem should be analyzed by using elastic shell theory. In the past decades, the subject was studied theoretically and experimentally, both for clamped-clamped and cantilevered shells.

Consider a cylindrical shell either conveying fluid or immersed in an axial flow, as shown in Figure 7. In this case, however, the internal flow can no longer be treated as a plug flow, but rather as a three-dimensional one. The linear equations of motion may be written as [86]

$$\ell_1(u,v,w) = \gamma \frac{\partial^2 u}{\partial t^2}, \qquad \ell_1(u,v,w) = \gamma \frac{\partial^2 v}{\partial t^2},$$

$$\ell_1(u,v,w) = -\gamma \left[\frac{\partial^2 w}{\partial t^2} - \frac{q_r}{\rho_s h} \right], \tag{7}$$

In which e_i (i = 1, 2, 3) denote linear differential operators of the axial coordinate x and the circumferential angle θ; u, v and w are defined, respectively, as the axial, circumferential and radial displacements of the middle surface of the shell; q_r is the radial surface loading per unit area, which can be written as

$$q_r = p_i - p_e. \tag{8}$$

In the above equation, p_i and p_e are the internal and external pressures exerted on the shell. The fluid is assumed to be inviscid and incompressible for simplicity; the flow is irrotational. Moreover, p_i and p_e are supposed to be composed of mean steady components and the perturbation components.

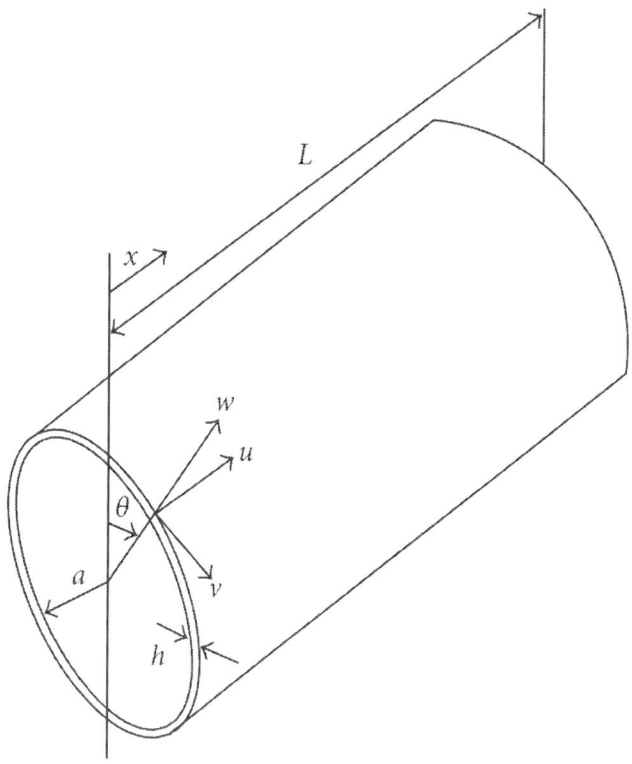

Figure 7: A cylindrical shell, showing some key dimensions.

If the effect of the steady components is ignored, the perturbation components may be obtained via potential flow theory [86] and can be written as

$$p_i^* = -\frac{\rho_i a}{n + \lambda I_{n+1}(\lambda)/I_n(\lambda)} \left[\frac{\partial}{\partial t} + U_i \frac{\partial}{\partial x}\right]^2 w,$$
(9)

$$p_e^* = -\frac{\rho_e a}{n - \lambda K_{n+1}(\lambda)/K_n(\lambda)} \left[\frac{\partial}{\partial t} + U_e \frac{\partial}{\partial x}\right]^2 w,$$
(10)

at r = a − 0+ and r = a + 0+, respectively, for h/a ≪ 1(h and an are defined in Figure 7). In the above two equations, ρ_i and ρ_e are the fluid densities of the internal and external flows, respectively; U_i and U_e are, respectively, defined as the internal and external flow velocities; λ and n are the axial wavenumber and the circumferential wavenumber, respectively. For the particular case of the internal or external fluid being quiescent, (9) and (10) still apply but with U_i = 0 or U_o = 0.

It can be seen that the terms arising from the squarebrackets operator in (9) and (10) can be written as $\partial^2 w/\partial t^2 + U^2 \partial^2 w/\partial x^2 + 2U\partial^2 w/\partial x \partial t$. These various terms are associated, respectively, with the inertia of the fluid, and the centrifugal and Coriolis forces of the moving fluid. Thus, the fluid effect is wholly analogous to that acting on a straight pipe conveying fluid. It is not surprising that the mechanism of underlying instabilities appears to be quite similar to that of beam-like instabilities of pipes conveying fluid. If the shell is supported at both ends, corresponding to a conservative system, it loses stability by divergence at a certain flow velocity. For a slightly higher flow velocity, the shell may subject to a coupled-mode flutter. Once again, the reliability of the post divergence flutter predicted by means of linear theory is questionable and hence needs to be re-examined by means of nonlinear theory, as discussed later.

In the case of a cantilevered shell conveying internal fluid, corresponding to a neoconservatives system, the instability is in the form of single-mode flutter, similar to the case of thicker pipes. The dynamics with external flow is quite similar to that with internal flow. This form of single-mode flutter is what was observed experimentally [85, 86].

Experiments with elastomer shells and air-flow showed that cantilevered shells lose stability by flutter, as predicted by theory [86, 87]. In the case of shells with clamped ends, however, the dynamics observed in experiments with internal and external flows were quite different: with external flow the system lost stability by divergence, while with internal flow the system lost stability by flutter [86, 88]. Therefore, the dynamics of shells conveying internal fluid predicted by linear theory did not agree with that observed experimentally, since divergence has not been detected in experiments. It would seem that re-examination of the dynamics was necessary. Clearly, this re-examination must involve both nonlinear theory and further experiments. This was discussed in greater detail by Karagiozis et al. [89–91] and Païdoussis [13].

The question of nonoccurrence of divergence in experiments with shells conveying internal flow was resolved [13, 92] by conducting experiments with stiffer shells. It was found that the stiffer shells did indeed lose stability by divergence, which is what is predicted by theory. Based on the nonlinear theory, Amabili et al. [93] further analyzed a shell with simply supported ends conveying fluid. They predicted that the shell would lose stability by divergence via a strongly subcritical pitchfork bifurcation. However, they did not develop coupled-model flutter. As the experiments [89] were always done with shells with clamped ends (for experimental convenience), a new nonlinear theoretical model was developed for shells with clamped ends [90, 91]. Again, the postdivergence flutter was not detected. Theory and experiments are in reasonable agreement with each other quantitatively.

DYNAMICS OF PLATES IN AXIAL FLOW

Vibration of flexible plates due to axial flow is an important issue, for instance, in paper manufacturing and paper printing [94, 95], also in parallel-plate assemblies used as core elements in some research and power nuclear reactors.

Solitary Plate in Axial Flow

First, consider a two dimensional plate of flexural rigidity $\wp = Eh^3/[12(1-\upsilon^2)]$, E and υ being the elastic modulus and Poisson ratio, respectively, h being the plate thickness and ρ_p the density. The plate is subjected to a flow-related perturbation pressure p∗. The schematic of the analytical system is shown in Figure 8. The linear equation of motion can be written as

$$\wp \frac{\partial^4 w}{\partial x^4} + C_d \frac{\partial w}{\partial t} + \rho_p h \frac{\partial^2 w}{\partial t^2} = -p^*, \tag{11}$$

where C_d is viscous damping coefficient. It is assumed that the flow is inviscid; the complete solution for p∗ has been obtained by Kornecki et al. [96] and is given by

$$-p^* = \frac{\rho}{\pi L}\left\{\int_0^1\left[L^2\frac{\partial^2 w}{\partial t^2} + 2UL\frac{\partial^2 w}{\partial \xi \partial t} + U^2\frac{\partial^2 w}{\partial \xi^2}\right]\right.$$
$$\left. \times \ln|\bar{x}-\xi|\,d\xi - R(\bar{x})\right\}, \tag{12}$$

where

$$R(\bar{x}) = U^2\left\{\left[w'(1) + (L/U)\dot{w}(1)\right]\ln(1-\bar{x})\right.$$
$$\left. -\left[w'(0) + (L/U)\dot{w}(0)\right]\ln\bar{x}\right\}, \tag{13}$$

in which $\bar{x}$ and ξ are defined by $\bar{x}= x/L$ and $\xi = u/L$, respectively, u being a dummy variable; t is dimensional time; ρ is the fluid density.

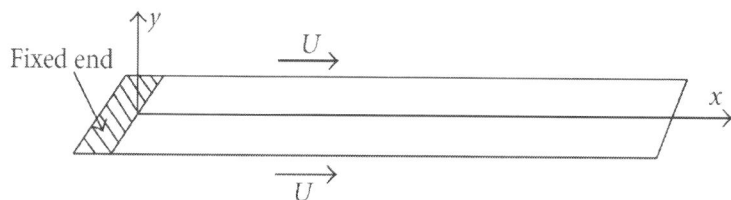

Figure 8: Schematic of a solitary plate in axial flow.

The terms given in the square-brackets on the righthand side of (12) have the similar functional form as in (1) and the function $\ln|\bar{x} - \xi|$ may be viewed as the effect of spatial memory. The difference in the theories mentioned in Sections 2, 3, 4, 5, and 6 and that for the plate here is significant. For pipes, tubular beams and shells of the local fluid forces depend only on the local displacement. For the plate problem, however, the local fluid forces depend on the global flow field.

By using linear theory, it has been found that a plate with supported ends loses stability by divergence and then by coupled-model flutter subsequently. These theoretical findings were broadly supported by experiments conducted by Dugundji et al. [97]. It is recalled that, for fluid-conveying pipes with both ends supported, the coupled-model flutter instability, which is predicted by the linear theory, has not been observed experimentally. In the case of supported plates subjected to axial flow, however, this coupled-model flutter was indeed observed in experiments.

For a cantilevered plate subjected to axial flow, the dynamics is much clearer; the system loses stability by flutter, as predicted by theory. Also, this form of instability has been observed experimentally. The similarity in the form of the fluttering plate to a fluttering pipe, in all its features, is quite remarkable. In the case of a cantilevered plate, however, the prediction is much more complicated, since the vorticity shed by the flapping plate into the wake should be taken into account [97–100].

More recently, Tang and Païdoussis [101–103] further investigated the instability and nonlinear vibrations of two-dimensional cantilevered plates in axial flow. In [101, 102], a nonlinear equation of motion has been utilized, assuming the middle plane of the plate to be inextensible, together with the unsteady lumped-vortex model for calculating the unsteady fluid loads. The flutter boundary obtained was compared with available experimental data. It was found that, when the plate is long, the theoretical predictions are in very good agreement with measurements from different experiments. In contrast, agreement with experiments is rather poor for short plates. In another recent work reported by Tang and Païdoussis [103], the nonlinear vibration of the same dynamical system developed in [101, 102] was further studied. However, an additional spring support of either linear or cubic type was installed at various locations on the plate. When the flow velocity is sufficiently high, the plate was predicted to exhibit chaotic motions via a period-doubling route. However, these interesting dynamical behaviours should be examined experimentally.

Parallel-Plate Assembly

A parallel-plate assembly, generally, consists of many thin plates stacked in parallel; between these parallel plates there are narrow channels to let coolant flow through. The main problem in this type of fuel system is the static and/or dynamic instabilities due to the flowing fluid. In some practical tests, large deflections and/or flutter were observed when the flow velocity became sufficiently high (see, e.g., [104–107]). The large deflections and/or flutter might lead to failure in practice.

Many attempts have been made to theoretically analyze the vibrations and study instability. Perhaps the first study on this problem was by Miller [108], who presented a theory for predicting the critical velocity for static divergence (collapse) of parallel-plate assembly. In his analysis, based on wide-beam theory and Bernoulli's theorem, the critical velocity was obtained by equating pressure differences between channels to the elastic restoring force of a plate. Miller's theory was further improved by Johansson [109], who included the effects of fluid friction and flow redistribution.

Scavuzzo [110] and Wambsganss [111] made further improvements upon Miller's and Johansson's model by considering the nonlinearity caused by large deflections (i.e., geometric nonlinearity). Rosenberg and Youngdahl [112] formulated a dynamical model and obtained the same critical velocity by using a two-dimensional mode. Yang and Zhang [113] developed a multispan elastic beam model to imitate a typical substructure of a parallel-plate structure. In their analytical model, there exists a narrow channel between the lower surface of the wide beam and the upper surface of the bottom plate of the water trough. By using the added water mass and damping coefficients, the free vibrational frequencies of the system were analyzed. Yang and Zhang [114] further investigated a parallel flat plate-type structure in rigid water trough or rigid rectangular tube.

More recently, Guo and Païdoussis [115] developed a more accurate and general theoretical analysis for parallel-plate assembly system. In their analysis, the plates were treated as two dimensional, with a finite length, and the flow field is taken to be inviscid, three dimensional. In [115], the equation of motion of an elastic plate is given by

$$\wp\left(\frac{\partial^4 w}{\partial x^4} + \frac{\partial^4 w}{\partial x^2 \partial y^2} + \frac{\partial^4 w}{\partial y^4}\right) + \rho_p h \frac{\partial^2 w}{\partial t^2} + P^* = 0, \tag{14}$$

where $P^* = P^*(x, y, t)$ may be viewed as the net load per unit area on the plate, equal to the difference between the perturbation pressures on the upper and lower surfaces of the plate caused by its deflection. Because of antisymmetry with respect to the plate, the perturbation pressures on the upper and lower surfaces must be equal in magnitude, but opposite in sign.

Based on (14), several important conclusions were [115] (i) single-mode divergence, mostly in the first mode, and coupled-mode flutter involving adjacent modes were found; (ii) the frequencies at a given flow velocity and the critical velocities increase as the aspect ratio decreases; (iii) in the case of large aspect ratios and small channel-height-to-plate-width ratios, the plates lose stability by first-mode divergence, however, very short plates usually lose stability by coupled-model flutter in the first and second modes; (iv) critical velocities for both divergence and flutter are insensitive to changes in damping coefficients.

Before closing this section, it should be remarked that most studies discussed in the foregoing were based on linear theories for parallel-plate assembly. Unfortunately, the literature on the nonlinear dynamics of such type of structures is very limited. If the essential nonlinearity is accounted for, the dynamical behaviour may be much richer.

DYNAMICS OF SLENDER STRUCTURES IN AXIAL FLOW OR AXIALLY TOWED IN QUIESCENT FLUID

Slender Structures in Axial Flow

As indicated in the Introduction, although most failures of slender structures (mostly cylinders or rods) are associated with the conditions of cross-flow, the cases of axial flow have also been shown to be of significance. In the first such study, motivated by application to the vibration of fissile fuel rods in nuclear reactors, Païdoussis [116] has investigated the dynamics of cylinders or rods in axial flow. In his two later papers, Païdoussis [117, 118] led to a still-used semi-empirical relation for predicting the turbulence-induced vibration levels in such systems.

Further research, however, was mainly driven by curiosity. Nevertheless, many applications can be found in practice. These applications should include, but not limited to (i) dynamics of rods and reactivity monitors in nuclear reactors; (ii) the vibration in closely spaced clusters of cylinders; (iii) the turbulence-induced vibration of tube (or cylinder) arrays in heat exchangers.

Solitary Cylinders or Rods

The schematic of a cantilevered cylinder in axial flow is shown in Figure 9. As developed in [119], the simplest form of the linear equation of motion of a cylinder in axial flow may be written as

$$EI\frac{\partial^4 w}{\partial x^4} + MU^2\frac{\partial^2 w}{\partial x^2} + 2MU\frac{\partial^2 w}{\partial x \partial t} + (m + M)\frac{\partial^2 w}{\partial t^2}$$
$$- \left[\frac{1}{2}\rho DU^2 C_T(L - x) + \frac{1}{2}\rho D^2 U^2 C_b\right]\frac{\partial^2 w}{\partial x^2}$$
$$+ \frac{1}{2}\rho DUC_N\left(\frac{\partial w}{\partial t} + U\frac{\partial w}{\partial x}\right) + \frac{1}{2}\rho DC_D\frac{\partial w}{\partial t} = 0, \quad (15)$$

where $M = \rho A$ is the virtual, or added, mass of the fluid per unit length for unconfined flow, A being the cross-sectional area of the cylinder and ρ the fluid density, w(x,t) is the lateral deflection, C_T and C_N are viscous force coefficients in the longitudinal and normal direction, respectively, C_D is the linearized zero-flow viscous drag coefficient for lateral motions, C_b is the base drag coefficient, D is the diameter of the cylinder, and the other symbols are the same as for internal flow.

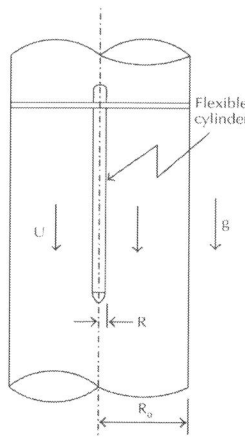

Figure 9: Schematic of a hanging cylinder in axial flow.

If both ends are supported, the equation of motion is slightly more complicated, depending also on whether the downstream end is free to slide axially, pressurization of the external fluid, and so on.

In the case of a cantilevered cylinder, it was generally supposed that the free-end is ogive-shaped. The simplest of the boundary conditions are

$$EI\frac{\partial^2 w}{\partial x^2} = 0,$$

$$EI\frac{\partial^3 w}{\partial x^3} + fMU\left(\frac{\partial w}{\partial t} + U\frac{\partial w}{\partial x}\right) - (m + fM)x_e\frac{\partial^2 w}{\partial t^2} = 0, \quad (16)$$

where $x_e = (1/A)\int_{L-l}^{L}(1/A)A(x)\,dx$, l being the length of the shaped end; f is a parameter first introduced by Hawthorne [120], equal to unity for a truly streamlined end. However, f is generally smaller because of 3D flow over the free-end and boundary-layer effects.

From (15), it is immediately seen that its first line is identical to the equation of motion of straight pipes conveying fluid (see (1)). Examining (15), it can be found that its second line is associated with the viscosity of the fluid. In fact, (1) and (15) differ only because of the viscous terms constituting in (15). Generally, the viscous terms are small compared to the first four terms in (15). Therefore, it may be expected that the dynamics, for cylinders with supported ends at least, is similar to that of the fluid-conveying pipe. This similarity is confirmed by Païdoussis [119].

For a cylinder with simply supported ends, as discussed in [119], divergence can be predicted at a no dimensional flow velocity only slightly higher than $\bar{U} = \pi$ in the first mode, followed at $\bar{U} \approx 2\pi$ by divergence in the second mode. Coupled-flutter is predicted at $\bar{U} \approx 6.48$. Indeed, postdivergence (couple-mode) flutter has been observed in experiments [121]. In fact, recent calculations by means of nonlinear theory have confirmed the existence of postdivergence flutter [122], and more recently reconfirmed experimentally [123]. In the study of [122], it was found that a Hopf bifurcation arises from loss of stability of the trivial equilibrium state. More interestingly, the system displays quasiperiodic and chaotic motions at higher flows, of course, predicted by means of nonlinear theory

The dynamics of a cantilevered system should be discussed here, since it is not similar to that of a cantilevered pipe conveying fluid. As reported in [119], for f = 0.8, the cantilevered system first loses stability by divergence at $\bar{U} \approx 2.04$, and then by single-mode flutter at $\bar{U} \approx 5.16$, and after destabilizations by flutter in the third mode at $\bar{U} \approx 8.17$. The reader may be surprised by the fact that the cantilevered system first loses stability by divergence at low-flow velocity. The reason is that the divergence is related to the presence of the tapered free-end. It is recalled that the case f = 0.8 suggests a fairly well-streamlined end. If f = 0, however, the end is blunt, and hence divergence is not possible.

The dynamics of cantilevered system predicted by linear theory has been re-examined theoretically, by means of nonlinear theory, and experimentally [124, 125]. It was found that the essential dynamical behaviour is as predicted by linear theory. However, the bifurcations do not arise in the same way.

Clustered Cylinders

The dynamics of clustered cylinders in axial flow has received considerable attention [126,127], because such systems exist in many engineering applications, as discussed in the foregoing.

The vibrations of such cylinders compared with that of isolated cylinder are characterized by (i) the effect of proximity of the other cylinders being important, causing various instabilities to occur at lower flow velocities, and (ii) the effect of intercylinder motion coupling, decreasing the critical flow velocities. Therefore, predicting the critical flow velocities for instability requires one very important piece of data: the cluster geometry and the intercylinder separation. For more details on this topic, one can refer to [127].

Slender Structures Towed in Quiescent Fluid

A considerable amount of notable work has been conducted on towed cylinders, rods, or tubular beams. Many applications of this system have emerged as follows: (i) vibrations of extruding metal and plastic rods in fluid [6, 7]; (ii) stability and vibrations of extremely long "seismic arrays," mostly towed behind boats and used in mineral exploration in the deep seas; (iii) the vibrations of towed pipelines for easy relocation mostly used in ocean; (iv) the vibrations of articulated submarine transporters; (v) high-speed trains traveling in narrow tunnels.

A sketch of towed slender structures shown in Figure 10 has received considerable attention [128, 129]. For this dynamical model, it was found that the dynamics with low towing speed is dominated by rigid-body instabilities. At higher towing speeds flexural instabilities might arise, much as for a cantilevered cylinder. Recently, in a work by Langre et al. [130], flutter was indeed found to exist.

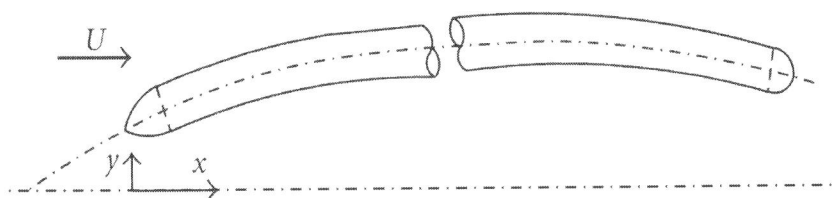

Figure 10: Idealized system of a towed cylinder with no cylindrical "nose" and "tail" segments.

Another typical system of a flexible cylinder or a tubular beam axially towed in fluid is shown in Figure 11. In Figure 11(a), a slender cantilevered beam is extending axially in the horizontal direction at a known rate, while immersed in a dense incompressible fluid. This tubular beam system has been studied by Taleb and Misra [6] and Gosselin et al. [7]. In the study by Gosselin et al., the fluid-dynamic forces obtained by Taleb and Misra were perceived to be not correctly accounted for. Thus, Gosselin et al. [7] re-examined the fluid-dynamic forces. It was found that, in the case of low constant extension rates, the system displays a phase of oscillation with increasing amplitude and decreasing frequency until the motion is strongly damped and later becomes statically unstable. For faster deployment rates, the beam has a short flutter phase at the beginning of the deployment, followed by a brief phase of damped oscillation until it exhibits static divergence. For fast enough deployment rates, the system is unstable from the beginning and never stabilizes. It should be mentioned that the effective length of the towed beam is increased with time, since the axially moving beam is clamped-free. It was also found that the axial added mass coefficient plays significant role in the stability of the system.

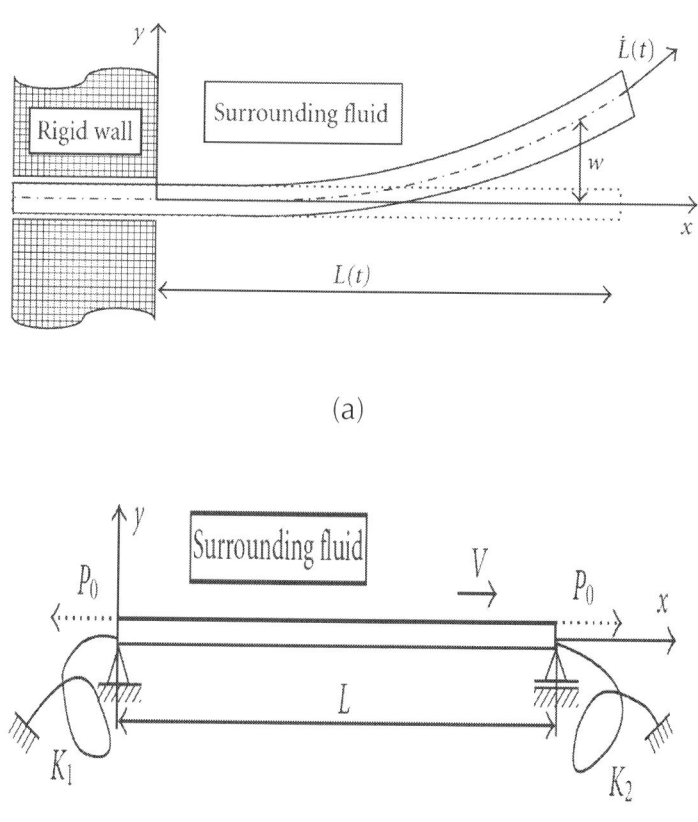

(a)

(b)

Figure 11: The cantilevered and supported axially moving beams. (a) The system considered in [7]. (b) The system considered in [8].

More recently, an axially towed system in fluid, shown in Figure 11(b), was investigated by Lin and Qiao [8]. Compared with the system in Gosselin et al. [7], this beam system has hybrid supports at both ends. It is worth noting that the formula of the total axial tension (T(x)) in the supported beam is different from that in the cantilevered system. In the study of Gosselin et al. [7], a nonzero value of T (L) arises from drag-induced compression at the freeend. Therefore, the final equation of motion of a supported system does differ from that of a cantilevered system. Compared with the cantilevered system, the effective length of the supported beam system keeps constant with time

Figure 12 [8] shows the effect of moving speed on the variation of the lowest three eigenvalues ($\Omega 1$, $\Omega 2$, and $\Omega 3$) of the moving beam with pinned-pined supports. It was found that the first mode first becomes unstable by divergence instability when the moving speed becomes equal or larger than the lowest critical moving speed (no dimensional) $\overline{V} \approx 1.06$. However, flutter instability was predicted to occur at a higher moving speed ($\overline{V} = 2.27$) in the first mode. The critical moving speeds at which divergence may occur in the second and third modes were both higher than $\overline{V} = 2.27$.

260   Dynamics of Offshore Structures

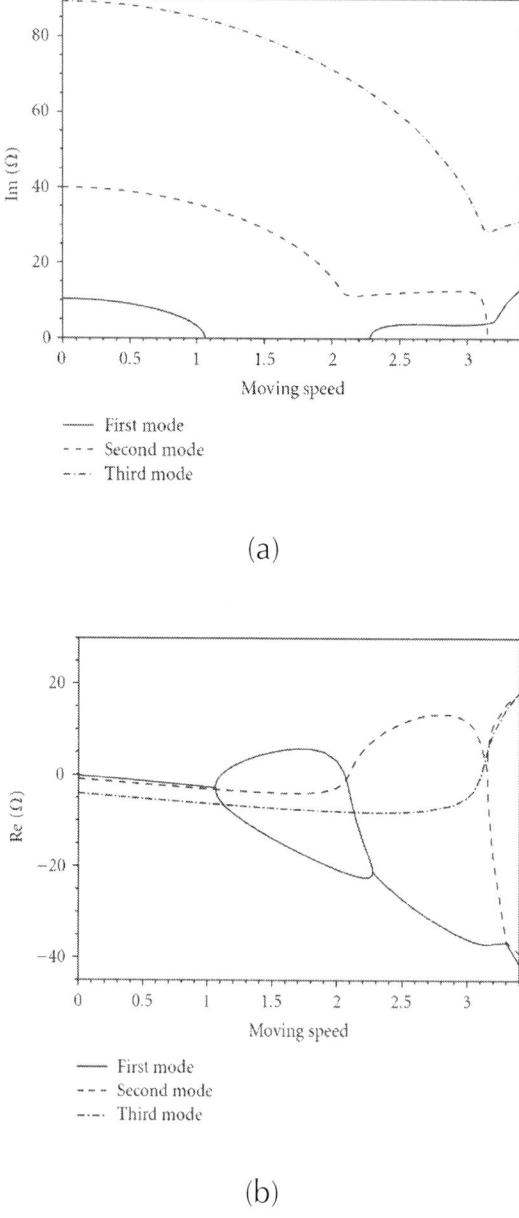

Figure 12: The imaginary and real components of the dimensionless frequency, Ω, as functions of the moving speed, $\bar{V}$, for the lowest three modes of a pinned-pinned beam; from [8].

It is of special interest to see the case of clamped-clamped moving beam. In this case, typical results are shown in Figure 13 [8]. It is obviously seen that divergence in the first mode occurs at $\bar{V} = 2.072$ and in the third at $\bar{V} = 4.18$. However, the second mode was predicted to be stable in the range $0 < \bar{V} < 4.3$.

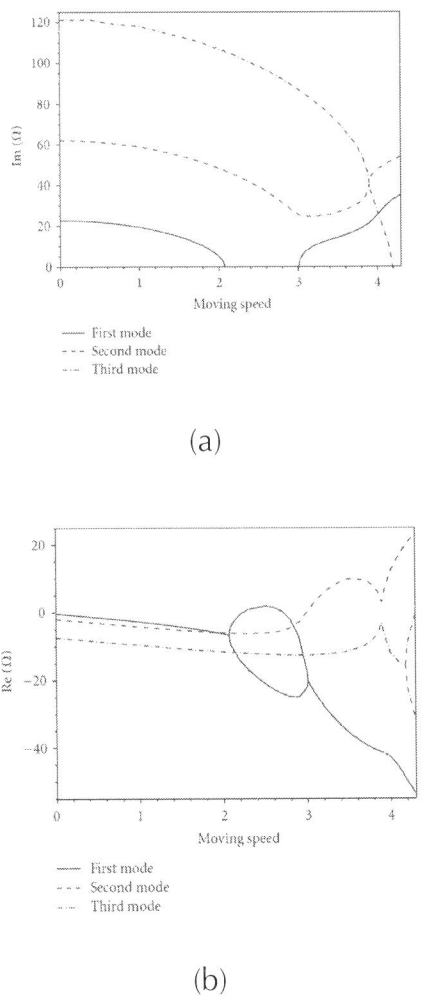

Figure 13: The imaginary and real components of the dimensionless frequency, Ω, as functions of the moving speed, $\bar{V}$ for the lowest three modes of a clamped-clamped beam; from [8].

CONCLUSIONS

The knowledge base associated with the basic dynamics of slender structures subjected to axial flow or towed in quiescent fluid has expanded greatly in recent years, as the number of applications continues to grow. Obviously, the literature survey is not exhaustive. For example, in Becker's survey [131] associated with the topic of pipes conveying fluid, 223 references were cited; in a more recent review by Païdoussis and Li [132], again, more than 200 references associated with pipes conveying fluid were cited. This paper, therefore, presents a selective review of the research undertaken on the vibrations of slender structures subjected to axial flow or axially towed in fluid. Undoubtedly many important contributions have been missed. Other aspects, which have been covered in a recent book by Païdoussis [133], also have not been fully discussed here.

One item of particular interest to the readers is that the fundamental understanding and experience gained and the methodology employed in the studies of the dynamical model of fluid-conveying pipes has proved to be very useful in the study of several other dynamical models, particularly shells conveying or immersed in axial flow, nanotubes conveying fluid, tubular beams subjected to both internal and external axial flows, cylinders and plates in axial flow, and tubular beams or cylinders axially towed in fluid. As reported in [132], it can be now understood that the dynamics of thin shells containing or surrounded by annular flows, with applications to aircraft engines and some types of nuclear reactors utilizing thermal shields. Similarly, the understanding of the dynamics of heat exchanger tube arrays and nuclear reactor fuel clusters subjected to external axial flows owes a great deal to the understanding gained in the study of pipes conveying internal flow. Furthermore, the dynamics of various cylindrical beam-like components subjected to internal or annular flows (e.g., solar thermal power plant chimney subjected to internal and external flows) can be understood in terms of what has been presented in the fluid-conveying thin pipe system. Interestingly, however, as indicated in [132], all these mostly unexpected applications came

10–40 years after the basic research on the topic had already been done.

From the history of investigating the slender structures subjected to axial flow or axially towed in fluid, it can be seen that, various important issues, such as, but not limited to, vibrations of nanotubes conveying fluid when accounting for small-length effect, instability of pipes aspirating fluid, and nonlinear vibrations of tubular beams concurrently subjected to internal and external axial flows, remain not wholly resolved. To the authors' knowledge, several items may be of future interest on analyzing the stability and vibrations of slender structures associated with axial-flow-forces.

- With the development of computer techniques, utilizing advanced numerical approaches to simulate the dynamical behaviour of slender structures subjected to axial flow or towed in quiescent fluid becomes realistic. To-day, the simulation results may play an important role, intuitively showing the dynamical behaviour of such systems. To advance in this direction, numerical (CFD) studies using ANSYS or other procedures should be initiated, which would help reveal the dynamics closer to the truth. The CFD results may meet the requirement for better coupling between the solid and fluid models, and making fluid models more realistic.

- The effect resulting from the nanoscale on the vibration of nanotubes conveying fluid has not been included so far. At small length scales the material micro-structure becomes increasingly important and its effect can no longer be ignored. Thus, the direct use of classic continuum approach to small length scales may be questionable. It is a possible solution to extend the classic continuum approach to smaller length scales by accounting for the information regarding the behavior of material microstructure. Therefore, some new theoretical models should be developed to resolve such an important issue.

- In practice, the axial flow is always with stochastic velocity or has a stochastic component superposed on steady flow. Therefore, the study on stochastic dynamics of slender

structures subjected to axial flow has more practical application in this area. However, the literature on this topic is quite limited.
- Nonlinear problems of various slender structures discussed in Sections 2, 3, 4, 5, 6, 7, and 8are not wholly resolved. For example, the nonlinear equation of motion of parallel-plate assembly in axial flow, slender structures axially towed in quiescent fluid, tubular beams subjected to both internal and external axial flows (may not be independent), and pipes aspirating fluid, have not been derived out, and hence the corresponding nonlinear dynamics has not been explored yet. Therefore, much attention may be concentrated on the nonlinear aspects of those slender structures mentioned in the foregoing.
- In the past decades, various methods of vibration control mostly considered the linear equations of motion for slender structures subjected to axial flow. More importantly, in a vibration control system, time-delayed feedback unavoidably exists. As reported by Xu and Chung [134], time delayed feedback may change the stability and dynamics of dynamical systems, leading to much more complex dynamical behaviors. To suppress the amplified oscillations, therefore, the methods of nonlinear control for slender structures, subjected to axial flows or towed in fluid, should also be developed by considering time delayed feedback.

ACKNOWLEDGMENTS

This work is supported by the National Natural Science Foundation of China (10772071 and 10802031) and the Scientific Research Foundation of HUST (2006Q003B).

REFERENCES

1. M. P. Païdoussis, "Flow-induced vibration in nuclear reactors and heat exchangers: practical experiences and state of knowledge," in Practical Experiences with Flow-Induced Vibrations, E. Naudascher and D. Rockwell, Eds., pp. 1–81, Springer, Berlin, Germany, 1980.

2. J. X. Xia, J. R. Ni, and C. Mendoza, "Hydraulic lifting of manganese nodules through a riser," Journal of Offshore Mechanics and Arctic Engineering, vol. 126, no. 1, pp. 72–77, 2004.

3. Y. A. Khulief, F. A. Al-Sulaiman, and S. Bashmal, "Vibration analysis of drillstrings with self-excited stick-slip oscillations," Journal of Sound and Vibration, vol. 299, no. 3, pp. 540–558, 2007.

4. M. P. Païdoussis, "Dynamics of flexible slender cylinders in axial flow—part 1: theory," Journal of Fluid Mechanics, vol. 26, no. 4, pp. 717–736, 1966.

5. M. P. Païdoussis, "Dynamics of flexible slender cylinders in axial flow—part 2: experiments," Journal of Fluid Mechanics, vol. 26, no. 4, pp. 737–751, 1966.

6. I. A. Taleb and A. K. Misra, "Dynamics of an axially moving beam submerged in a fluid," Journal of Hydronautics, vol. 15, no. 1–4, pp. 62–66, 1981.

7. F. Gosselin, M. P. Païdoussis, and A. K. Misra, "Stability of a deploying/extruding beam in dense fluid," Journal of Sound and Vibration, vol. 299, no. 1-2, pp. 123–142, 2007.

8. W. Lin and N. Qiao, "Vibration and stability of an axially moving beam immersed in fluid," International Journal of Solids and Structures, vol. 45, no. 5, pp. 1445–1457, 2008.

9. M. P. Païdoussis and N. T. Issid, "Dynamic stability of pipes conveying fluid," Journal of Sound and Vibration, vol. 33, no. 3, pp. 267–294, 1974.

10. R. W. Gregory and M. P. Païdoussis, "Unstable oscillation of tubular cantilevers conveying fluid—I: theory," Proceedings

of the Royal Society of London. Series A, vol. 293, no. 1435, pp. 512–527, 1966.
11. G. L. Kuiper and A. V. Metrikine, "Dynamic stability of a submerged, free-hanging riser conveying fluid," Journal of Sound and Vibration, vol. 280, no. 3–5, pp. 1051–1065, 2005.
12. G. L. Kuiper and A. V. Metrikine, "Experimental investigation of dynamic stability of a cantilever pipe aspirating fluid," Journal of Fluids and Structures, vol. 24, no. 4, pp. 541–558, 2008. 16 Advances in Acoustics and Vibration
13. M. P. Païdoussis, "Some unresolved issues in fluid-structure interactions," Journal of Fluids and Structures, vol. 20, no. 6, pp. 871–890, 2005.
14. M. P. Païdoussis, C. Semler, and M. Wadham-Gagnon, "A reappraisal of why aspirating pipes do not flutter at infinitesimal flow," Journal of Fluids and Structures, vol. 20, no. 1, pp. 147–156, 2005.
15. G. T. S. Done and A. Simpson, "Dynamic stability of certain conservative and non-conservative systems," Journal of Mechanical Engineering Science, vol. 19, pp. 251–263, 1977.
16. P. J. Holmes, "Bifurcations to divergence and flutter in flowinduced oscillations: a finite dimensional analysis," Journal of Sound and Vibration, vol. 53, no. 4, pp. 471–503, 1977.
17. P. J. Holmes, "Pipes supported at both ends cannot flutter," Journal of Applied Mechanics, vol. 45, no. 3, pp. 619–622, 1978.
18. P. J. Holmes and J. Marsden, "Bifurcation to divergence and flutter in flow-induced oscillations: an infinite dimensional analysis," Automatica, vol. 14, no. 4, pp. 367–384, 1978.
19. L. Wang, "A further study on the non-linear dynamics of simply supported pipes conveying pulsating fluid," International Journal of Non-Linear Mechanics, vol. 44, no. 1, pp. 115–121, 2009.

20. L. N. Panda and R. C. Kar, "Nonlinear dynamics of a pipe conveying pulsating fluid with parametric and internal resonances," Nonlinear Dynamics, vol. 49, no. 1-2, pp. 9–30, 2007.
21. A. K. Bajaj, P. R. Sethna, and T. S. Lundgren, "Hopf bifurcation phenomena in tubes carrying a fluid," SIAM Journal on Applied Mathematics, vol. 39, no. 2, pp. 213–230, 1980.
22. A. K. Bajaj and P. R. Sethna, "Effect of symmetry-breaking perturbations on flow-induced oscillations in tubes," Journal of Fluids and Structures, vol. 5, no. 6, pp. 651–679, 1991.
23. D. M. Tang and E. H. Dowell, "Chaotic oscillations of a cantilevered pipe conveying fluid," Journal of Fluids and Structures, vol. 2, no. 3, pp. 263–283, 1988.
24. M. P. Païdoussis and F. C. Moon, "Nonlinear and chaotic fluidelastic vibrations of a flexible pipe conveying fluid," Journal of Fluids and Structures, vol. 2, no. 6, pp. 567–591, 1988.
25. M. P. Païdoussis, G. X. Li, and F. C. Moon, "Chaotic oscillations of the autonomous system of a constrained pipe conveying fluid," Journal of Sound and Vibration, vol. 135, no. 1, pp. 1–19, 1989.
26. M. P. Païdoussis, G. X. Li, and R. H. Rand, "Chaotic motions of a constrained pipe conveying fluid. comparison between simulation, analysis, and experiment," Journal of Applied Mechanics, vol. 58, no. 2, pp. 559–565, 1991.
27. M. P. Païdoussis and C. Semler, "Nonlinear and chaotic oscillations of a constrained cantilevered pipe conveying fluid: a full nonlinear analysis," Nonlinear Dynamics, vol. 4, no. 6, pp. 655–670, 1993.
28. M. P. Païdoussis, J. P. Cusumano, and G. S. Copeland, "Lowdimensional chaos in a flexible tube conveying fluid," Journal of Applied Mechanics, vol. 59, no. 1, pp. 196–205, 1992.

29. L. Wang and Q. Ni, "A note on the stability and chaotic motions of a restrained pipe conveying fluid," Journal of Sound and Vibration, vol. 296, no. 4-5, pp. 1079–1083, 2006.
30. G. S. Copeland and F. C. Moon, "Chaotic flow-induced vibration of a flexible tube with end mass," Journal of Fluids and Structures, vol. 6, no. 6, pp. 705–718, 1992.
31. M. P. Païdoussis, C. Semler, M. Wadham-Gagnon, and S. Saaid, "Dynamics of cantilevered pipes conveying fluid—part 2: dynamics of the system with intermediate spring support," Journal of Fluids and Structures, vol. 23, no. 4, pp. 569–587, 2007.
32. Y. Modarres-Sadeghi, C. Semler, M. Wadham-Gagnon, and M. P. Païdoussis, "Dynamics of cantilevered pipes conveying fluid—part 3: three-dimensional dynamics in the presence of an end-mass," Journal of Fluids and Structures, vol. 23, no. 4, pp. 589–603, 2007.
33. V. A. Svetlitskii, "Statics, stability and small vibrations of the flexible tubes conveying ideal incompressible fluid," Raschety na Prochnost, vol. 14, pp. 332–351, 1969.
34. V. A. Svetlitskii, "Vibration of tubes conveying fluids," The Journal of the Acoustical Society of America, vol. 62, no. 3, pp. 595–600, 1977.
35. V. A. Svetlitskii, Mekhanika Truboprovodov I Shlangov, Machinostronye, Moscow, Russia, 1982.
36. S.-S. Chen, "Vibration and stability of a uniformly curved tube conveying fluid," The Journal of the Acoustical Society of America, vol. 51, no. 1B, pp. 223–232, 1972.
37. S.-S. Chen, "Flow-induced in-plane instabilities of curved pipes," Nuclear Engineering and Design, vol. 23, no. 1, pp. 29–38, 1972.
38. S.-S. Chen, "Out-of-plane vibration and stability of curved tubes conveying fluid," Journal of Applied Mechanics, vol. 40, no. 2, pp. 362–368, 1973.
39. R. W. Doll and C. D. Mote Jr., "The dynamic formulation and the finite element analysis of curved and twisted

tubes transporting fluids," Report to the National Science Foundation, Department of Mechanical Engineering, University of California, Berkeley, Calif, USA, 1974.

40. R. W. Doll and C. D. Mote Jr., "On the dynamic analysis of curved and twisted cylinders transporting fluids," Journal of Pressure Vessel Technology, vol. 98, no. 2, pp. 143–150, 1976.

41. J. L. Hill and C. G. Davis, "The effect of initial forces on the hydroelastic vibration and stability of planar curved tubes," Journal of Applied Mechanics, vol. 41, no. 2, pp. 355–359, 1974.

42. C. Dupuis and J. Rousselet, "Application of the transfer matrix method to non-conservative systems involving fluid flow in curved pipes," Journal of Sound and Vibration, vol. 98, no. 3, pp. 415–429, 1985.

43. A. K. Misra, M. P. Païdoussis, and K. S. Van, "On the dynamics of curved pipes transporting fluid—part I: inextensible theory," Journal of Fluids and Structures, vol. 2, no. 3, pp. 221– 244, 1988.

44. A. K. Misra, M. P. Païdoussis, and K. S. Van, "On the dynamics of curved pipes transporting fluid—part II: extensible theory," Journal of Fluids and Structures, vol. 2, no. 3, pp. 245– 261, 1988.

45. A. K. Misra, M. P. Païdoussis, and K. S. Van, "Dynamics and stability of fluid conveying curved pipes," in Proceedings of the International Symposium on Flow-Induced Vibration and Noise, vol. 4, pp. 1–24, ASME, Chicago, Ill, USA, NovemberDecember 1988.

46. Q. Ni, L. Wang, and Q. Qian, "Chaotic transients in a curved fluid conveying tube," Acta Mechanica Solida Sinica, vol. 18, no. 3, pp. 207–214, 2005.

47. N. Qiao, W. Lin, and Q. Qin, "Bifurcations and chaotic motions of a curved pipe conveying fluid with nonlinear constraints," Computers and Structures, vol. 84, no. 10-11, pp. 708–717, 2006. Advances in Acoustics and Vibration 17

48. W. Lin, N. Qiao, and H. Yuying, "Dynamical behaviors of a fluid-conveying curved pipe subjected to motion constraints and harmonic excitation," Journal of Sound and Vibration, vol. 306, no. 3-5, pp. 955–967, 2007.
49. D. Jung and J. Chung, "In-plane and out-of-plane motions of an extensible semi-circular pipe conveying fluid," Journal of Sound and Vibration, vol. 311, no. 1-2, pp. 408–420, 2008.
50. Y. Huang, G. Zeng, and F. Wei, "A new matrix method for solving vibration and stability of curved pipes conveying fluid," Journal of Sound and Vibration, vol. 251, no. 2, pp. 215–225, 2002.
51. Q. Ni and Y. Huang, "Differential quadrature method to stability analysis of pipes conveying fluid with spring support," Acta Mechanica Solida Sinica, vol. 13, no. 4, pp. 320–327, 2000.
52. W. Lin and N. Qiao, "In-plane vibration analyses of curved pipes conveying fluid using the generalized differential quadrature rule," Computers and Structures, vol. 86, no. 1-2, pp. 133–139, 2008.
53. C. Dupuis and J. Rousselet, "The equations of motion of curved pipes conveying fluid," Journal of Sound and Vibration, vol. 153, no. 3, pp. 473–489, 1992.
54. S. Iijima, "Helical microtubules of graphitic carbon," Nature, vol. 354, no. 6348, pp. 56–58, 1991.
55. R. F. Gibson, E. O. Ayorinde, and Y.-F. Wen, "Vibrations of carbon nanotubes and their composites: a review," Composites Science and Technology, vol. 67, no. 1, pp. 1–28, 2007.
56. G. Hummer, J. C. Rasaiah, and J. P. Noworyta, "Water conduction through the hydrophobic channel of a carbon nanotube," Nature, vol. 414, no. 6860, pp. 188–190, 2001.
57. A. Karlsson, R. Karlsson, M. Karlsson, et al., "Molecular engineering: networks of nanotubes and containers," Nature, vol. 409, no. 6817, pp. 150–152, 2001.
58. Y. Gao and Y. Bando, "Carbon nanothermometer containing gallium," Nature, vol. 415, no. 6872, p. 599, 2002.

59. Z. Mao and S. B. Sinnott, "A computational study of molecular diffusion and dynamic flow through carbon nanotubes," The Journal of Physical Chemistry B, vol. 104, no. 19, pp. 4618–4624, 2000.
60. Y. Gogotsi, J. A. Libera, A. Guvenc ̈ ̧-Yazicioglu, and C. M. Megaridis, "In situ multiphase fluid experiments in hydrothermal carbon nanotubes," Applied Physics Letters, vol. 79, no. 7, pp. 1021–1023, 2001.
61. V. P. Sokhan, D. Nicholson, and N. Quirke, "Fluid flow in nanopores: accurate boundary conditions for carbon nanotubes," The Journal of Chemical Physics, vol. 117, no. 18, pp. 8531–8539, 2002.
62. Y. Liu, Q. Wang, T. Wu, and L. Zhang, "Fluid structure and transport properties of water inside carbon nanotubes," The Journal of Chemical Physics, vol. 123, no. 23, Article ID 234701, 7 pages, 2005.
63. R. E. Tuzun, D. W. Noid, B. G. Sumpter, and R. C. Merkle, "Dynamics of fluid flow inside carbon nanotubes," Nanotechnology, vol. 7, no. 3, pp. 241–246, 1996.
64. T. Natsuki, Q.-Q. Ni, and M. Endo, "Wave propagation in single- and double-walled carbon nanotubes filled with fluids," Journal of Applied Physics, vol. 101, no. 3, Article ID 034319, 5 pages, 2007.
65. J. Yoon, C. Q. Ru, and A. Mioduchowski, "Vibration and instability of carbon nanotubes conveying fluid," Composites Science and Technology, vol. 65, no. 9, pp. 1326–1336, 2005.
66. J. Yoon, C. Q. Ru, and A. Mioduchowski, "Flow-induced flutter instability of cantilever carbon nanotubes," International Journal of Solids and Structures, vol. 43, no. 11-12, pp. 3337–3349, 2006.
67. C. D. Reddy, C. Lu, S. Rajendran, and K. M. Liew, "Free vibration analysis of fluid-conveying single-walled carbon nanotubes," Applied Physics Letters, vol. 90, no. 13, Article ID 133122, 3 pages, 2007.

68. X. Wang, X. Y. Wang, and G. G. Sheng, "The coupling vibration of fluid-filled carbon nanotubes," Journal of Physics D, vol. 40, no. 8, pp. 2563–2572, 2007.
69. L. Wang and Q. Ni, "On vibration and instability of carbon nanotubes conveying fluid," Computational Materials Science, vol. 43, no. 2, pp. 399–402, 2008.
70. L. Wang, Q. Ni, M. Li, and Q. Qian, "The thermal effect on vibration and instability of carbon nanotubes conveying fluid," Physica E, vol. 40, no. 10, pp. 3179–3182, 2008.
71. Y. Yan, W. Q. Wang, and L. X. Zhang, "Dynamical behaviors of fluid-conveyed multi-walled carbon nanotubes," Applied Mathematical Modelling, vol. 33, no. 3, pp. 1430–1440, 2009.
72. L. Wang, Q. Ni, and M. Li, "Buckling instability of doublewall carbon nanotubes conveying fluid," Computational Materials Science, vol. 44, no. 2, pp. 821–825, 2008.
73. Y. Yan, X. Q. He, L. X. Zhang, and C. M. Wang, "Dynamic behavior of triple-walled carbon nanotubes conveying fluid," Journal of Sound and Vibration, vol. 319, no. 3–5, pp. 1003–1018, 2009.
74. K. Dong, B. Y. Liu, and X. Wang, "Wave propagation in fluid-filled multi-walled carbon nanotubes embedded in elastic matrix," Computational Materials Science, vol. 42, no. 1, pp. 139–148, 2008.
75. J. P. Den Hartog, "John Orr memorial lecture: recent cases of mechanical vibration," The South African Mechanical Engineer, vol. 19, no. 3, pp. 53–68, 1969.
76. J. J. Bailey and I. Finnie, "An analytical study of drillstring vibration," Journal of Engineering for Industry, vol. 82, no. 2, pp. 122–128, 1960.
77. R. W. Tucker and C. Wang, "An integrated model for drillstring dynamics," Journal of Sound and Vibration, vol. 224, no. 1, pp. 123–165, 1999.
78. F. Cesari and S. Curioni, "Buckling instability in tubes subject to internal and external axial fluid flow," in Proceedings of the

4th Conference on Dimensioning, pp. 301–311, Hungarian Academy of Science, Budapest, Hungary, October 1971.

79. M. J. Hannoyer and M. P. Païdoussis, "Instabilities of tubular beams simultaneously subjected to internal and external axial flows," Journal of Mechanical Design, vol. 100, pp. 328–336, 1978.

80. J. V. Grigoriev, "Stability of a drill tube column with an initial curvature in the axial stream," Journal of Bauman Moscow State Technical University: Mashinostronye, vol. 5, pp. 23–28, 1978 (Russian).

81. M. P. Païdoussis and P. Besancon, "Dynamics of arrays of cylinders with internal and external axial flow," Journal of Sound and Vibration, vol. 76, no. 3, pp. 361–379, 1981.

82. X. Wang and F. Bloom, "Dynamics of a submerged and inclined concentric pipe system with internal and external flows," Journal of Fluids and Structures, vol. 13, no. 4, pp. 443–460, 1999.

83. M. P. Païdoussis, T. P. Luu, and S. Prabhakar, "Dynamics of a long tubular cantilever conveying fluid downwards, which then flows upwards around the cantilever as a confined annular flow," Journal of Fluids and Structures, vol. 24, no. 1, pp. 111–128, 2008. 18 Advances in Acoustics and Vibration

84. T. P. Luu, On the dynamics of three systems involving tubular beams conveying fluid, M.Eng. thesis, Department of Mechanical Engineering, McGill University, Montreal, Canada, 1983.

85. M. P. Païdoussis and J.-P. Denise, "Flutter of cylindrical shells conveying fluid," Journal of Sound and Vibration, vol. 16, pp. 456–461, 1971.

86. M. P. Païdoussis and J.-P. Denise, "Flutter of thin cylindrical shells conveying fluid," Journal of Sound and Vibration, vol. 20, no. 1, pp. 9–26, 1972.

87. V. B. Nguyen, M. P. Païdoussis, and A. K. Misra, "An experimental study of the stability of cantilevered coaxial

cylindrical shells conveying fluid," Journal of Fluids and Structures, vol. 7, no. 8, pp. 913–930, 1993.

88. A. El Chebair, M. P. Païdoussis, and A. K. Misra, "Experimental study of annular flow-induced instabilities of cylindrical shells," Journal of Fluids and Structures, vol. 3, no. 4, pp. 349–364, 1989.

89. K. N. Karagiozis, M. P. Païdoussis, A. K. Misra, and E. Grinevich, "An experimental study of the nonlinear dynamics of cylindrical shells with clamped ends subjected to axial flow," Journal of Fluids and Structures, vol. 20, no. 6, pp. 801–816, 2005.

90. K. N. Karagiozis, M. P. Païdoussis, M. Amabili, and A. K. Misra, "Nonlinear stability of cylindrical shells subjected to axial flow: theory and experiments," Journal of Sound and Vibration, vol. 309, no. 3–5, pp. 637–676, 2008.

91. K. N. Karagiozis, M. P. Païdoussis, and A. K. Misra, "Transmural pressure effects on the stability of clamped cylindrical shells subjected to internal fluid flow: theory and experiments," International Journal of Non-Linear Mechanics, vol. 42, no. 1, pp. 13–23, 2007.

92. K. N. Karagiozis, M. P. Païdoussis, E. Grinevich, A. K. Misra, and M. Amabili, "Stability and nonlinear dynamics of clamped circular cylindrical shells in contact with flowing fluid," in Proceedings of IUTAM Symposium on Integrated Modeling of Fully Coupled Fluid Structure Interactions Using Analysis, Computations and Experiments, pp. 375–390, Kluwer Academic Publishers, New Brunswick, NJ, USA, June 2003.

93. M. Amabili, F. Pellicano, and M. P. Païdoussis, "Nonlinear dynamics and stability of circular cylindrical shells containing flowing fluid—I: stability," Journal of Sound and Vibration, vol. 225, no. 4, pp. 655–699, 1999.

94. Y. Watanabe, S. Suzuki, M. Sugihara, and Y. Sueoka, "An experimental study of paper flutter," Journal of Fluids and Structures, vol. 16, no. 4, pp. 529–542, 2002.

95. C. Lemaitre, P. Hemon, and E. de Langre, "Instability of a long ribbon hanging in axial air flow," Journal of Fluids and Structures, vol. 20, no. 7, pp. 913–925, 2005.
96. A. Kornecki, E. H. Dowell, and J. O'Brien, "On the aeroelastic instability of two-dimensional panels in uniform incompressible flow," Journal of Sound and Vibration, vol. 47, no. 2, pp. 163–178, 1976.
97. J. Dugundji, E. H. Dowell, and B. Perkins, "Subsonic flutter of panels on continuous elastic foundations," AIAA Journal, vol. 1, no. 5, pp. 1146–1154, 1963.
98. T. Ishii, "Aeroelastic instabilities of simply supported panels in subsonic flow," in American Institute of Aeronautics and Astronautics, Royal Aeronautical Society, and Japan Society for Aeronautical and Space Sciences, Aircraft Design and Technology Meeting, Los Angeles, Calif, USA, November 1965, Paper 65-772.
99. D. S. Weaver and T. E. Unny, "Hydroelastic stability of a flat plate," Journal of Applied Mechanics, vol. 37, no. 3, pp. 823–827, 1970.
100. C. H. Ellen, "Stability of simply supported rectangular surfaces in uniform subsonic flow," Journal of Applied Mechanics, vol. 40, no. 1, pp. 68–72, 1973.
101. L. Tang and M. P. Païdoussis, "On the instability and the postcritical behaviour of two-dimensional cantilevered flexible plates in axial flow," Journal of Sound and Vibration, vol. 305, no. 1-2, pp. 97–115, 2007.
102. L. Tang and M. P. Païdoussis, "The influence of the wake on the stability of cantilevered flexible plates in axial flow," Journal of Sound and Vibration, vol. 310, no. 3, pp. 512–526, 2008.
103. L. Tang and M. P. Païdoussis, "The dynamics of two-dimensional cantilevered plates with an additional spring support in axial flow," Nonlinear Dynamics, vol. 51, no. 3, pp. 429–438, 2008.

104. W. K. Doan, "The engineering test reactor-a status report," Nucleonics, vol. 16, no. 1, pp. 102–105, 1958.
105. W. L. Zabriskie, "An experimental evaluation of the effect of length-to-width ratio on the critical flow velocity of single plate assemblies," Tech. Rep. 59GL209, General Electric Company, General Engineering Laboratory, Schenectady, NY, USA, 1959.
106. R. D. Groninger and J. J. Kane, "Flow induced deflections of parallel flat plates," Nuclear Science and Engineering, vol. 16, pp. 218–226, 1963.
107. G. E. Smissaert, "Static and dynamic hydro-elastic instabilities in MTR-type fuel elements—part I: introduction and experimental investigation," Nuclear Engineering and Design, vol. 7, no. 6, pp. 535–546, 1968.
108. D. R. Miller, "Critical flow velocities for collapse of reactor parallel-plate fuel assemblies," Journal of Engineering for Power, vol. 82, pp. 83–95, 1960.
109. R. B. Johansson, "Hydraulic instability of reactor parallel plate fuel assemblies," in Nuclear Engineering Science Conference, New York, NY, USA, April 1960, preprint paper no. 57.
110. R. J. Scavuzzo, "Hydraulic instability of flat parallel-plate assemblies," Nuclear Science and Engineering, vol. 21, pp. 463–472, 1965.
111. M. W. Wambsganss Jr., "Second-order effects as related to critical coolant flow velocities and reactor parallel plate fuel assemblies," Nuclear Engineering and Design, vol. 5, no. 3, pp. 268–276, 1967.
112. G. S. Rosenberg and C. K. Youngdahl, "A simplified dynamic model for the vibration frequencies and critical coolant flow velocities for reactor parallel plate fuel assemblies," Nuclear Science and Engineering, vol. 13, pp. 91–102, 1962.
113. Y.-R. Yang and J.-Y. Zhang, "Frequency analysis of a parallel flat plate-type structure in still water—part I: a multi-span beam," Journal of Sound and Vibration, vol. 203, no. 5, pp. 795–804, 1997.

114. Y.-R. Yang and J.-Y. Zhang, "Frequency analysis of a parallel flat plate-type structure in still water—part II: a complex structure," Journal of Sound and Vibration, vol. 203, no. 5, pp. 805–814, 1997.
115. C. Q. Guo and M. P. Païdoussis, "Analysis of hydroelastic instabilities of rectangular parallel-plate assemblies," Journal of Pressure Vessel Technology, vol. 122, no. 4, pp. 171–176, 2000.
116. M. P. Païdoussis, "The amplitude of fluid-induced vibration of cylinders in axial flow," Tech. Rep. AECL-2225, Atomic Energy of Canada, Ontario, Canada, 1965. Advances in Acoustics and Vibration 19
117. M. P. Païdoussis, "An experimental study of vibration of flexible cylinders induced by nominally axial flow," Nuclear Science and Engineering, vol. 35, pp. 127–138, 1969.
118. M. P. Païdoussis, "Vibrations of cylindrical structures subjected to axial flow," Journal of Engineering for Industry, vol. 96, pp. 547–552, 1974.
119. M. P. Païdoussis, "Dynamics of cylindrical structures subjected to axial flow," Journal of Sound and Vibration, vol. 29, no. 3, pp. 365–385, 1973.
120. W. R. Hawthorne, "The early development of the Dracone flexible barge," Proceedings of the Institution of Mechanical Engineers, vol. 175, pp. 52–83, 1961.
121. M. P. Païdoussis, "Dynamics of flexible slender cylinders in axial flow—part 2: experiments," Journal of Fluid Mechanics, vol. 26, no. 4, pp. 737–751, 1966.
122. Y. Modarres-Sadeghi, M. P. Païdoussis, and C. Semler, "A nonlinear model for an extensible slender flexible cylinder subjected to axial flow," Journal of Fluids and Structures, vol. 21, no. 5–7, pp. 609–627, 2005.
123. Y. Modarres-Sadeghi, M. P. Païdoussis, C. Semler, and E. Grinevich, "Experiments on vertical slender flexible cylinders clamped at both ends and subjected to axial flow,"

Philosophical Transactions of the Royal Society A, vol. 366, no. 1868, pp. 1275–1296, 2008.

124. M. P. Païdoussis, E. Grinevich, D. Adamovic, and C. Semler, "Linear and nonlinear dynamics of cantilevered cylinders in axial flow—part 1: physical dynamics," Journal of Fluids and Structures, vol. 16, no. 6, pp. 691–713, 2002.

125. C. Semler, J. L. Lopes, N. Augu, and M. P. Païdoussis, "Linear and nonlinear dynamics of cantilevered cylinders in axial flow—part 3: nonlinear dynamics," Journal of Fluids and Structures, vol. 16, no. 6, pp. 739–759, 2002.

126. M. P. Païdoussis, "The dynamics of clusters of flexible cylinders in axial flow: theory and experiments," Journal of Sound and Vibration, vol. 65, no. 3, pp. 391–417, 1979.

127. S. S. Chen, Flow-Induced Vibration of Circular Structures, Hemisphere, Washington, DC, USA, 1987.

128. M. P. Païdoussis, "Stability of towed, totally submerged flexible cylinders," Journal of Fluid Mechanics, vol. 34, no. 2, pp. 273–297, 1968.

129. A. P. Dowling, "The dynamics of towed flexible cylinders—part 1: neutrally buoyant elements," Journal of Fluid Mechanics, vol. 187, pp. 507–532, 1988.

130. E. de Langre, M. P. Païdoussis, O. Doare, and Y. Modarres-Sadeghi, "Flutter of long flexible cylinders in axial flow," Journal of Fluid Mechanics, vol. 571, pp. 371–389, 2007.

131. O. Becker, "Das durchstromte Rohr—Literaturbericht," Tech. Rep. IHZ-M80-212, der Ingenieurhochschule Zittau, Zittau, Germany, 1981.

132. M. P. Païdoussis and G. X. Li, "Pipes conveying fluid: a model dynamical problem," Journal of Fluids and Structures, vol. 7, no. 2, pp. 137–204, 1993.

133. M. P. Païdoussis, Fluid-Structure Interactions: Slender Structures and Axial Flow. Volume 2, Academic Press, London, UK, 2004.

134. J. Xu and K. W. Chung, "Effects of time delayed position feedback on a van der Pol-Duffing oscillator," Physica D, vol. 180, no. 1-2, pp. 17–39, 2003.

Chapter 9

Water-Air CO$_2$ Fluxes in the Tagus Estuary Plume (Portugal) during Two Distinct Winter Episodes

Ana P Oliveira[1], Marcos D Mateus[2], Graça Cabeçadas[1], and Ramiro Neves[2]

[1]Instituto Português do Mar e da Atmosfera (IPMA), I.P., Avenida de Brasília, Lisboa, 1449-006, Portugal

[2]MARETEC, Instituto Superior Técnico, Universidade de Lisboa, Av. Rovisco Pais, Lisboa, 1049-001, Portugal

ABSTRACT

Background

Estuarine plumes are frequently under strong influence of land-derived inputs of organic matter. These plumes have characteristic physical and chemical conditions, and their morphology and extent in the coastal area depends strongly on physical conditions such as river discharge, tides and wind action. In this work we investigate the physical dynamics of the Tagus estuary plume and the CO_2 system response during two contrasting hydrological winter periods. A hydrodynamic model was used to simulate the circulation regime of the study area, thus providing relevant information on hydrodynamic processes controlling the plume.

Results

Model simulations show that for the studied periods, the major cause of the plume variability (size and shape) was the interaction between Tagus River discharge and wind. The freshwater intrusion on Tagus shelf exerted considerable influence on biochemical dynamics, allowing identification of two regions: a high nutrient region enriched in CO_2 inside the estuarine plume and another warmer region rich in phytoplankton in the outer plume.

Conclusion

The Tagus estuarine plume behaved as a weak source of CO_2 to the atmosphere, with estimated fluxes of 3.5 ± 3.7 and 27.0 ± 3.8 mmol C m^{-2} d^{-1} for February 2004 and March 2001, respectively.

BACKGROUND

Coastal regions are significantly influenced by land-derived discharges emanating from estuaries, with estuarine plumes mediating the fluxes of natural terrestrial compounds and pollutant into shelf seas [1], [2]. The extent and morphology of estuarine plumes are a direct consequence of river discharge, but are also strongly dependent on other physical conditions such as tide and wind stress.

An essential characteristic of estuarine plumes may be defined by a significant salinity gradient, although the boundary of the plume is often difficult to define given the highly dynamic nature of such systems [1]. Furthermore, highly stratified plumes lead to well defined density fronts along their boundaries, where turbidity is relatively low and chlorophyll a relatively high, even in winter [3].

Some studies concerning nutrients, fluxes of organic constituents and phytoplankton have been undertaken in estuaries and/or their associated plumes [1],[4]-[8]; some highlight the seasonality CO_2 source/sink behaviour of the estuarine plumes [9]-[12], and only a few refer the estuarine plume dynamics and the carbonate system response [13],[14]. However, the CO_2 uptake capacity of the estuarine plumes in several continental shelf zones is already extensively reported [2], [3], [15]-[20], suggesting that other estuarine plumes might counteract inner estuary CO_2 emissions. The processes controlling the CO_2 dynamics in the estuarine plume are linked to various factors such as spring/summer phytoplankton blooms, thermodynamic effects, winter floods from the inner estuary or stratification/mixing of the plume water column. On an annual basis, these processes together with the complexity of near shore ecosystems, can significantly impact water-air CO_2 exchanges in estuarine plumes [13], [14].

Significant drawdown of CO_2 partial pressure (pCO_2), biological uptake of dissolved inorganic carbon (DIC) and an associated enhancement of dissolved oxygen and pH within plumes occur due to enhanced biological activity, as reported for the Mississippi River plume in the USA [21],[22], the Scheldt plume in Belgium [23], and the Pearl River estuary in China [8]. For the Changjiang Estuary plume (China), the pCO_2 drawdown and DO enhancement in the warm seasons (from April to October) appeared to be controlled by primary productivity and water-air exchange, while mixing dominated the aqueous pCO_2 in the cold seasons extending from November to March of the following year [15]. Mixing of river water with Gulf of Maine waters as also been pointed as responsible for the carbon variability in this system [24], although biological processes were significantly intense during the spring and summer seasons. Biological activity also lowers Amazon River plume pCO_2, and contributes to a CO_2 deficit in the northern western tropical North Atlantic Ocean that outlasts the plume's physical structure [25].

This paper aims to characterize the dynamics of water-air CO_2 flux in the Tagus estuarine plume (Figure 1) during two contrasting winter periods, based on the pCO_2 dynamics derived from field data. Underlying controlling mechanisms have been investigated based on the river discharge, the role of temperature and the biological activity. This study merges field data retrieved by experimental methods with information derived from the results of a numerical model on the spatial and temporal variability of the physical structure of the plume.

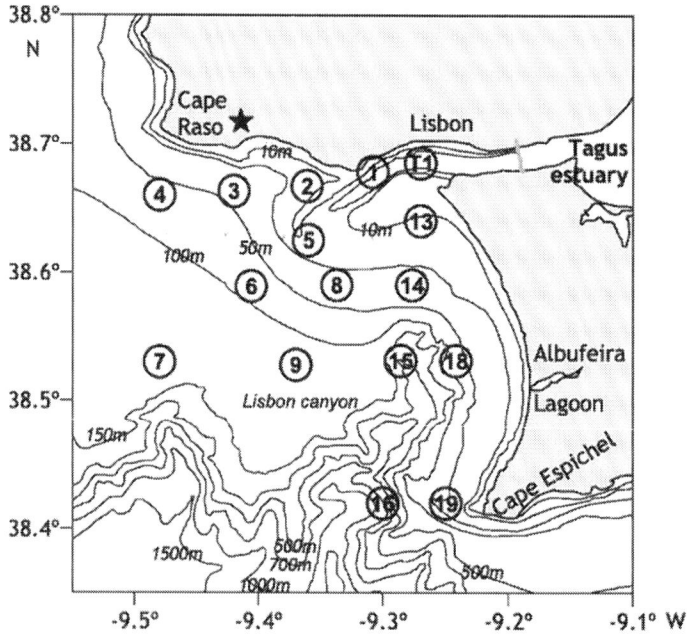

Figure 1: Location of the study site Location of the sampling stations in the mouth of the Tagus estuary (SW Portugal) and adjacent coastal area. The position of the Guia meteorological station (38°41′27″ N, 9°27′34″ W) is marked with a star.

RESULTS AND DISCUSSION

The sampling programs were carried out in winter, from 7 to 19 March 2001 and from 5 to 9 February 2004.

Environmental Settings

The first three months of 2001, with mean air temperature of 13.0°C, were slightly warmer than the same period in 2004, with mean air temperature of 11.5°C. The winter 2001 was characterised by exceptional rain events, with precipitation values significantly higher than during the same period in 2004. The effect of the different rainy regimes is seen in the Tagus flow, with a mean value of 1893 m^{3}s^{-1} in March 2001 and 481 m^3 s^{-1} in February 2004 (Figure 2). Atmospheric CO_2 ($pCO_{2,air}$) was slightly lower (mean value of 373 µatm) in 2001, when compared with 2004 (mean value of 380 µatm). Both periods were characterised by absence of upwelling, seen in the positive Bakun index mean values of 725 m^3 s^{-1} km^{-1} and of 344 m^3 s^{-1} km^{-1} for March 2001 and February 2004, respectively. Significant shifts in wind direction and intensity were observed in March 2001 (Figure 3A), with dominant direction from the SW quadrant and intensities between 7 – 10 m s^{-1}. In February 2004 the Tagus coastal area was under the influence of persistent south winds followed by stronger north winds (7 – 10 m s^{-1} in intensity), as shown in the wind rose in Figure 3B.

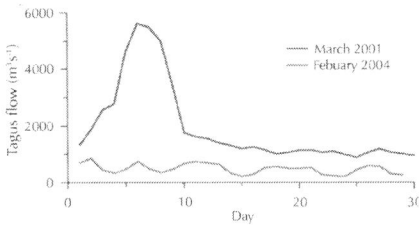

Figure 2: Tagus river outflow. Tagus river flow (m^3 s^{-1}) measured at a hydrometric station located upstream in March 2001 and February 2004. Values imposed in the modelled scenarios.

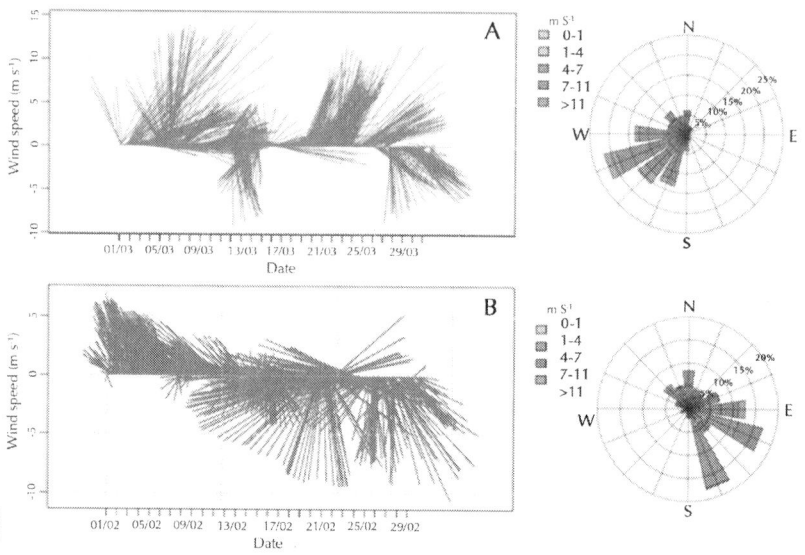

Figure 3: Wind regime. Stick diagram and wind rose of the wind regime measured at Guia meteorological station (38°41′27″ N, 9°27′34″ W) during (A) March 2001 and (B) February 2004. The wind intensity and direction showed here was used to force the model in each scenario. Wind rose shows the cumulative frequency in which wind speeds increase from the center to the outside.

The winter periods were considered statistically different (t-test, $p < 0.05$, n = 27) for all physical (T, S) and biogeochemical parameters (Si(OH)$_4$, AOU, SPM, Chl a, pH, TA, pCO_2), except for DO, NO_3, NH_4 and PO_4 (Table 1). Higher values for all parameters occurred in March 2001, except for S and pH, denoting the influence of the river plume. Salinity differences were also a consequence of the river flow in the two periods.

Table 1: Mean values for monitored parameters

	MARCH 2001		FEBRUARY 2004	
	Range	Mean value (SD[a])	Range	Mean value (SD[b])
T (°C)	14.8 – 15.9	15.4 (0.3)	14.4 – 14.9	14.7 (0.2)

S	31.4 – 34.6	32.3 (0.8)	27.9 – 35.7	32.7 (2.1)
NO_3 (µmol l^{-1})	0.9 – 13.7	8.4 (3.5)	1.1 – 15.8	8.4 (5.9)
NH_4 (µmol l^{-1})	0.5 – 3.3	1.9 (1.0)	0.4 – 4.4	2.6 (1.5)
PO_4 (µmol l^{-1})	0.2 – 1.0	0.5 (0.2)	0.2 – 1.1	0.7 (0.3)
$Si(OH)_4$ (µmol l^{-1})	4.8 – 26.5	17.1 (7.1)	1.1 – 16.9	9.6 (6.1)
DO (mg l^{-1})	8.3 – 8.9	8.5 (0.2)	7.6 – 9.3	8.2 (0.6)
AOU (µmol kg^{-1})	−25.9 – −1.7	−9.3 (6.8)	−35.6 – 25.8	2.0 (20.4)
SPM (mg l^{-1})	3.0 – 19.2	7.2 (4.6)	2.1 – 9.6	4.1 (2.2)
Chl a (mg m^{-3})	0.7 – 1.6	1.1 (0.3)	0.2 – 1.1	0.7 (0.3)
pH	7.88 – 7.96	7.93 (0.02)	8.06 – 8.17	8.10 (0.05)
TA (µmol kg^{-1})	3026 – 3770	3519 (282)	2357 – 2767	2489 (103)
pCO_2 (µatm)	990 – 1467	1207 (140)	431 – 654	531 (70)
Wind speed (m s^{-1})	3.4 – 3.7	-	1.0 – 4.2	-
Piston velocity (cm h^{-1})	3.32 – 4.18	-	0.04 – 5.54	-

[a] standard deviation (SD) (n = 13).

[b] standard deviation (SD) (n = 14).

Seawater surface range of data and mean values for Tagus coastal area during winter 2001 and 2004. Shaded area indicates the parameters that are statistically different (t-test, $p < 0.05$, n = 27) between the 2001 and 2004 winter sampling periods.

Oliveira et al.

Oliveira *et al. Carbon Balance and Management* 2015 10:2.

TA values in March 2001 are considerably high, but fall within empirically established boundaries. They are within the range reported for Tagus estuary adjacent coastal waters in previous works [26], [27]. Also, the Portuguese National Information System for Hydric Resources – SNIRH (data available at http://snirh.apambiente.pt/) reports TA values within the range 3000–8200 µmol kg^{-1} in the lower part of the estuary under the influence of freshwater. TA values in 2001 winter can be attributed to carbonate dissolution, which is confirmed by the significant decrease of particulate inorganic carbon from the estuary mouth (station T1, see

Figure 1) to the plume. Although anaerobic degradation processes, such as denitrification and sulphate reduction, can also impact alkalinity increase, there are no evidences that such processes occurred during the March 2001 sampling period. Carbon loads to the plume were also quite different in both winter periods, being the value in the 2001 winter (2931 t C d^{-1}) ~2.2 times higher than the value in 2004 winter (1340 t C d^{-1}).

The Estuarine Plume Boundary

The boundary of Tagus plume can be inferred by the salinity gradient resulting from the fresh water intrusion in the coastal area. As such, the size and extension of the plume is strongly related with riverine discharges and, thus, with rainfall. This is observed in the studied periods (Figure 4). Using the salinity isopleth 34.5 to set up the limit of the plume, it is possible to notice a larger plume in March 2001 as a result of higher river flow. During this period the plume is more pronounced, extending south to Albufeira Lagoon reaching the Espichel Cape limit (Figure 4A), ~30 km from the estuary mouth. In February 2004, the plume remains closer to the Tagus mouth, extending ~14 km north-west along the coast (Figure 4B). The T-S (Figure 4C, D) and AOU-S (Figure 4E, F) diagrams reflect the impact of Tagus water input on the coastal area adjacent to the estuary in terms of salinity, temperature and oxygen. During March 2001 the Tagus Bay was under the influence of Tagus discharge, as noticed by the salinity values below 34.5, temperature higher than 15°C, and by the oxygen super-saturated water (AOU < 0) (Figure 4C, E).

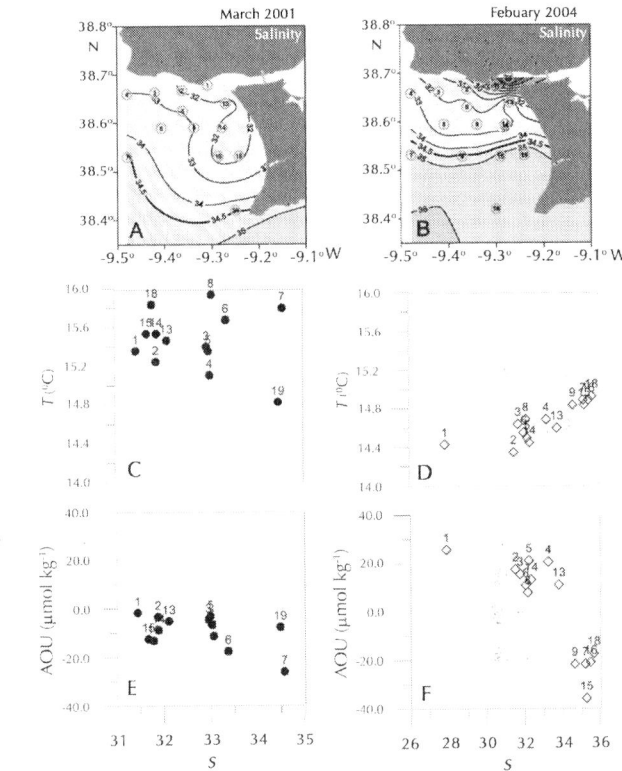

Figure 4: Tagus plume characterization. Surface salinity distribution during (A) March 2001 and (B) February 2004. T-S diagrams for (C) March 2001 and (D) February 2004 samplings, illustrating stations at the estuarine plume. AOU-S diagrams in (E) March 2001 and (F) February 2004 illustrating stations at the estuarine plume. The plume limit is represented by the 34.5 isopleth (bold line).

Model results show a significant variation in Tagus plume dispersion pattern between 20 and 30 March, as seen in Figure 5A and B. During this period, the horizontal current structure of the plume changes the northwest direction from the river mouth to the south. From 26 to 31 March the wind is consistently from the Northern quadrant with a relatively high intensity (~5 m s^{-1}) (Figure 3), inducing marked offshore/southward advection of the estuary plume. Current velocity intensifies as a result of the river flow increase in this period. Model results for the salinity provide insights on the plume size and shape (Figure 5B), and show significant variation in its limits in response to the wind regime. It is also noticed an evolution from its original position trapped along the northern side of the river mouth, in 23 March, to a south-west transport off the Tagus mouth.

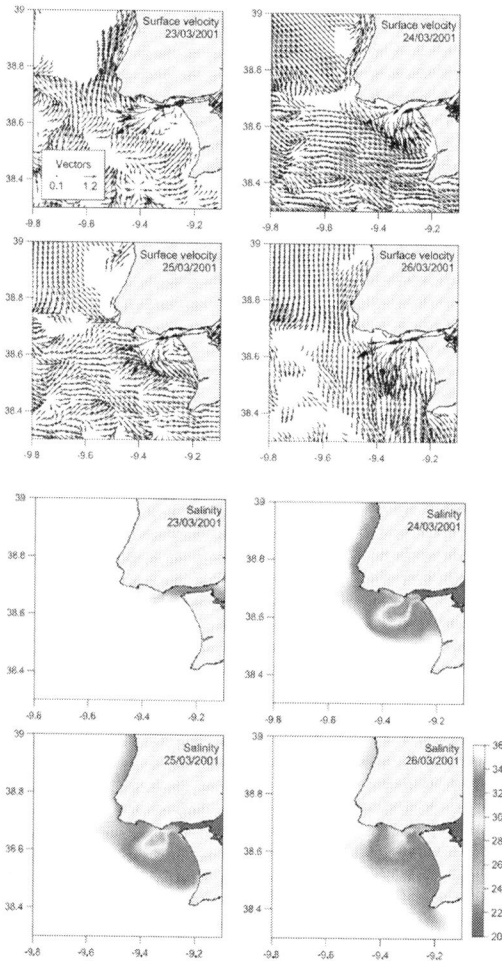

Figure 5: Model results. Model results for the surface currents and salinity in March 2001.

The 2004 winter period was characterised by water temperatures below 15°C (Figure 4D). Oxygen saturation showed the presence of the undersaturated plume (Stations 1 to 6, 8, 13 and 14), and an outer oversaturated area (Stations 7, 9, 15, 16 and 18) (Figure 4F). This is reinforced by the calculated DIC *versus* TA plot (Figure 6), where the separation of the two water masses (riverine and oceanic) is observed. Station 1 (with S < 30) was clearly isolated from other stations near the northern coast, displaying salinities between 30 and 34.5 (Figure 4D, F).

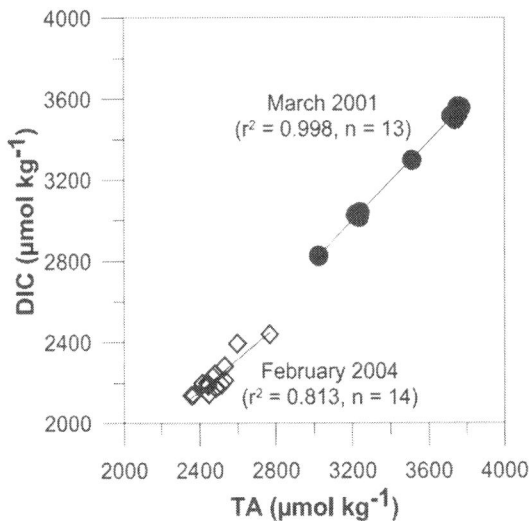

Figure 6: DIC vs. TA Calculated dissolved inorganic carbon (DIC) *versus* total alkalinity (TA) in Tagus coastal area during March 2001 and February 2004.

Simulated conditions for February 2004 show the formation of an estuary plume in the vicinity of the estuary mouth, extending westwards along the north side from the estuary as a result of geostrophic adjustment (Figure 7). This pattern is the result of the influence of moderate river flow and persistent south winds (Figure 3). Given the small variation in the forcing conditions, model results for February 2004 show little variation during the simulated period. The physical structure of the plume was consistently characterized by an offshore transport westward from the river mouth (Figure 7A), and presented a similar signature pattern in the horizontal salinity field observed in this period (Figure 7B).

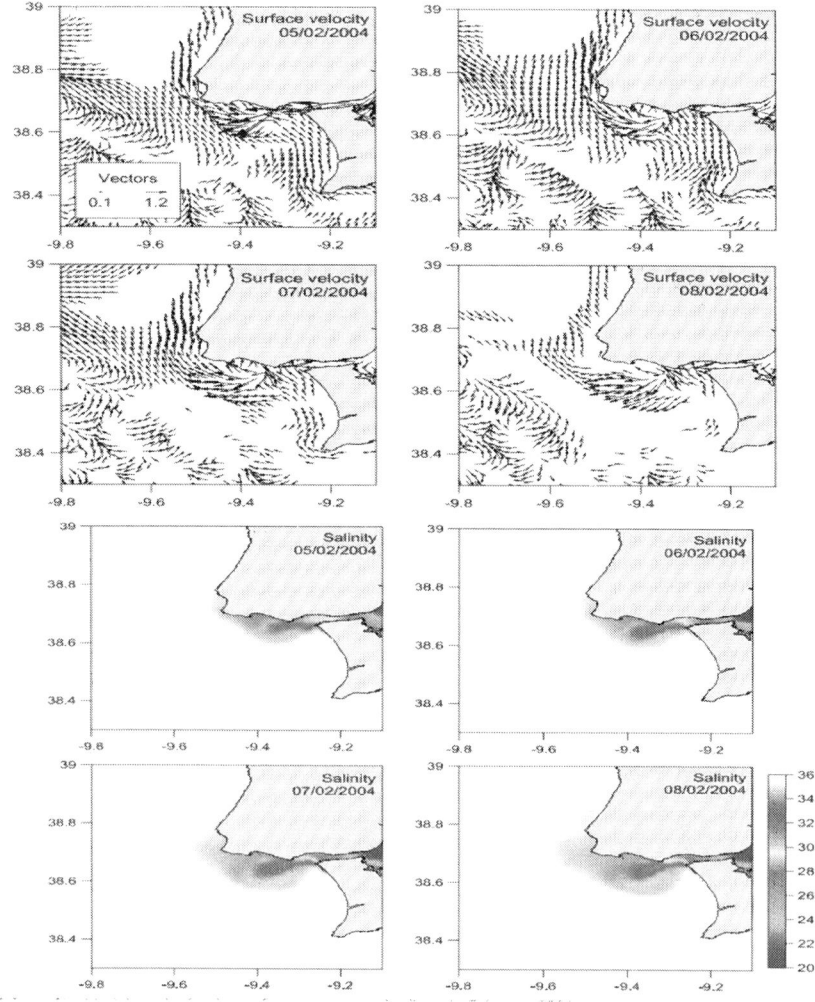

Figure 7: Model results. Model results for the surface currents and salinity in February 2004.

The signature of the estuarine plume is also evident in the chemical and biological parameters plotted in Figure 8. The results suggest a southward transport of the plume in winter 2001 and a northward transport in winter 2004. Model results explain these patterns by providing the temporal evolution of the physical conditions of the plume in both circumstances.

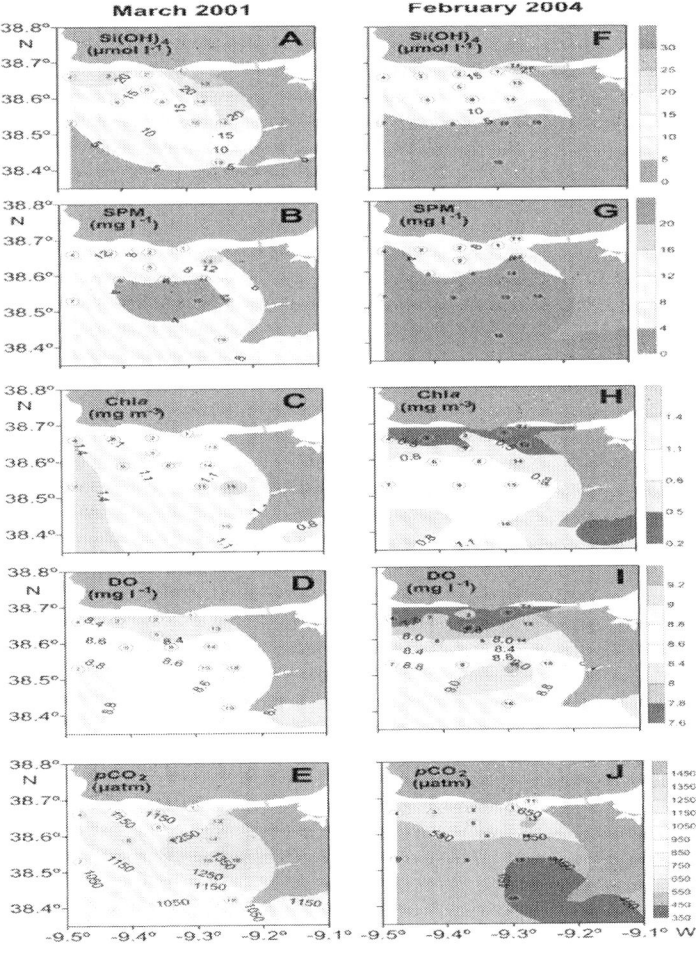

Figure 8: Surface concentration of monitored properties. Characterization of the Tagus coastal area during March 2001 and February 2004. Surface concentration of (A, F) silicate (Si (OH) 4), (B, G) suspended matter (SPM), (C, H) chlorophyll a (Chl a), (D, I) dissolved oxygen (DO), and (E, J) calculated CO_2 partial pressure (pCO_2).

The Estuarine Plume Biogeochemistry

Contour plots show a marked estuarine plume enriched in nutrients, represented as $Si(OH)_4$, particles and calculated pCO_2 in 2001 (Figure 8A, B, E). In both sampling periods, high concentrations of suspended material (SPM) were associated with low concentrations of particulate organic carbon (data not shown), suggesting an organically impoverished plume, similar to other estuarine plumes [6],[7],[28]. SPM declined outside the plume, partly due to sinking, and there was a nutrient decrease caused by the combined effects of mixing with nutrient poor offshore waters and phytoplankton uptake, as suggested by the increase in Chl a (Figure 8C, H) and DO (Figure 8D, I).

The different water masses observed in both scenarios are characterized by distinct environmental properties, and also reveal particular CO_2 features. Higher calculated pCO_2 occur in March 2001 (Table 1), ranging from 990 µatm outside the plume (Station 19) to 1460 µatm near Albufeira Lagoon (Station 18) in the tip of the Lisbon submarine canyon head (Figure 8E). The elevated pCO_2 values are probably a signal from the river where the most riverine station (~65 km of Station 1) values were up to 2500 µatm. In February 2004 the plume is pushed northwards (Figure 8J), and the highest value of calculated pCO_2 (654 µatm) is observed at the estuary mouth (Station 1). Both periods were characterised by CO_2 oversaturation, reaching ~400% in 2001 and ~170% in 2004. Other European estuarine plumes also presented CO_2 oversaturation, such as the Scheldt in winter [9],[12], the Elbe in the spring [6], and the Loire in autumn [5]. The marked pCO_2 gradient in the buoyant plume suggests that its structure and dynamics regulates the pCO_2 property in the studied area (Figure 8E, J). Again, several other studies revealed the variability of estuarine plumes with respect to CO_2 dynamics [29].

Calculated pCO_2 values decreased from inshore to offshore (Figure 8E, J), following the decreased in salinity. A significant correlation was found between calculated pCO_2 and salinity in February 2004 ($r^2 = 0.890$, $p < 0.05$, $n = 14$; Figure 9A). This distribution pattern, also seen in other systems[2],[9],[15],[25], indicates that the mixing processes influence the CO_2 pattern, which is reinforced by the proximity to the conservative mixing line (Figure 9A). However, the simultaneous calculated pCO_2 and DO decrease ($r^2 = 0.561$, $p < 0.05$, $n = 14$) and Chl a increases along the salinity gradient (Figure 9B, C), suggests the prominence of biological processes inside the plume. This is supported by the drawdown drop of calculated DIC associated with a pH increase ($r^2 = 0.704$, $p < 0.05$, $n = 14$) (Figure 9E, F). The non-linear relationships between calculated DIC and nutrients (only represented by NO_3 in this study) also reflect both mixing and biological processes (Figure 9G) regulating pCO_2 distribution in February 2004. These features were also found in the Mississippi River [22] and Amazon River [25] plumes.

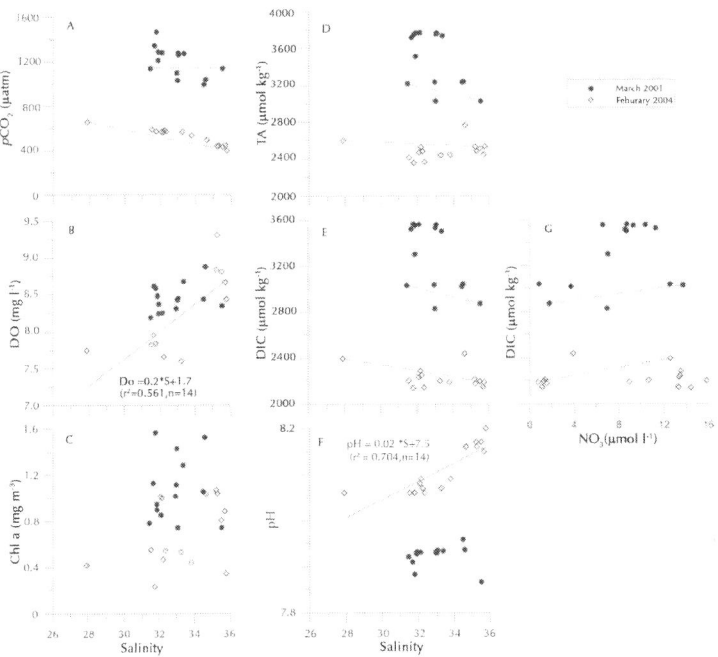

Figure 9: Correlations between parameters. Distributions of (A)calculated CO_2 partial pressure (pCO_2), (B) dissolved oxygen (DO), (C)chlorophyll a (Chl a), (D) total alkalinity (TA), (E) calculated dissolved inorganic car-

bon (DIC), and (F) pH along the salinity gradient, and of (G)DIC *versus* NO_3 for Tagus coastal area during March 2001 and February 2004. End-member mixing line is represented by the dotted line.

For the outer Loire estuary [13] biological processes, namely an episodic winter phytoplankton blooms, was also pointed out as responsible for pCO_2 variability. Applying the Takahashi et al. [30] procedure to the February 2004 data, pCO_2 variability was not affected by temperature. By contrast, in March 2001 pCO_2 variability was attributed to physical processes, such as the thermodynamic effect of temperature and the riverine/estuarine discharge [31]. Moreover, the non-linear relationships of TA and DIC with salinity (Figure 9D, E) and of calculated DIC and NO_3 (Figure 9G) reflect mixing and biological processes, inducing pCO_2 variability. However, the calculated DIC *vs.* TA plot suggests that DIC and TA resulted by the same mechanism.

For the outer Loire estuary [13] biological processes, namely an episodic winter phytoplankton blooms, was also pointed out as responsible for pCO_2 variability. Applying the Takahashi et al. [30] procedure to the February 2004 data, pCO_2 variability was not affected by temperature. By contrast, in March 2001 pCO_2 variability was attributed to physical processes, such as the thermodynamic effect of temperature and the riverine/estuarine discharge [31]. Moreover, the non-linear relationships of TA and DIC with salinity (Figure 9D, E) and of calculated DIC and NO_3 (Figure 9G) reflect mixing and biological processes, inducing pCO_2 variability. However, the calculated DIC *vs.* TA plot suggests that DIC and TA resulted by the same mechanism.

In both winter occasions two major regions were spatially individualized in the area: a high nutrient and CO_2 enriched region inside the plume, and a warmer region characterized by higher phytoplankton biomass in the outer plume.

CO_2 Fluxes Across the Water-air Interface

The water-air CO_2 fluxes showed similar patterns in the studied periods, with lower emissions to the atmosphere outside the plume (Table 2), and the highest values coincident with high wind speeds (data not shown). A striking pattern is found in both cases, namely the reduction in CO_2 fluxes from inside the plume to outside of about 90% and 20%, in 2001 and 2004, respectively. Other authors [2], [15]-[20] have also reported this CO_2 uptake capacity of estuarine plumes, even suggesting that other estuarine plumes might counteract inner estuary CO_2 emissions. Overall, the adjacent waters to the Tagus estuary acted as sources of CO_2 to the atmosphere, emitting 25.9 ± 4.3 mmol C m^{-2} d^{-1} in March 2001, and 2.4 ± 3.4 mmol C m^{-2} d^{-1} in February 2004 (Table 2). Thus, CO_2 emissions to the atmosphere in March 2001 were ~90% higher than in February 2004. The differences can be attributed to the variable river influence (e.g., effect of nutrients and labile organic matter, additional buoyant stability induced by freshwater fluxes), as suggested by other authors [10], [25],[32]. The CO_2 emissions estimated in this work are within the range of those reported for the Tagus adjacent coastal waters [31] and several other near-shore ecosystems [9].

Table 2: Water-air CO2 fluxes (Mean values and standard deviation)

	March 2001	February 2004
Estuarine plume (S < 34.5)	27.0 ± 3.8	3.5 ± 3.7
Outer plume (S > 34.5)	19.9 ± 1.0	0.2 ± 0.1
Overall area	25.9 ± 4.3	2.4 ± 3.4

Mean values and standard deviation of water-air CO_2 fluxes (mmol C m^{-2} d^{-1}) calculated according to [32] parameterization for stations inside and outside the Tagus estuary plume during March 2001 and February 2004.

Oliveira et al.

Oliveira *et al. Carbon Balance and Management* 2015 10:2.

CONCLUSIONS

Tagus estuarine plume can be traced on the shelf by gradients of salinity, but also by gradients of less conservative tracers such as water temperature, chlorophyll a, inorganic nutrients, total alkalinity and CO_2. Thus, Tagus estuary adjacent shelf exhibits a high nutrient, low chlorophyll and enriched in CO_2 estuarine plume and a warm region impoverished in CO_2 and enriched in phytoplankton in the outer plume. Estuarine Tagus plume behaved as a weak source of CO_2 to the atmosphere, with estimated fluxes of 3.5 ± 3.7 and 27.0 ± 3.8 mmol C m^{-2} d^{-1} for February 2004 and March 2001, respectively. Based on two winter cruises, it seems that Tagus plume significantly impacted estimates of water-air CO_2 fluxes at a regional scale. Hence, this work emphasizes the importance of estuarine plumes on the CO_2 dynamics in coastal areas. However, due to the complexity of near shore ecosystems and processes therein the magnitude of water-air fluxes is variable from one system to another. Also, this study reinforces the usefulness of complimentary approaches such as the application of numeric models in reproducing the physical and chemical characteristics of plumes dynamics. The model results provide the temporal evolution of the plume under varying wind and rivers discharge, providing additional information that could not be obtained otherwise and, consequently, insightful clues on the integration of field data. Still, this is a first approach to using modelling tools with field data in the Tagus estuary, and future developments will include the CO_2 dynamics in the model simulations.

METHODS

Study Area

The present investigation was carried out in the continental shelf offshore Tagus estuary (Figure 1) in the Portuguese coast, covering the geographic area between 38.35° – 38.80° N and 9.10° – 9.50° W. The continental shelf is ≤10 km wide south of Lisbon and presents topographic structures as prominent capes, promontories and submarine canyons. Its morphology is strongly influenced by the intense discharge of Tagus River, usually showing a pronounced dry/wet season signal as well as large inter-annual variation. The mean annual average discharge of Tagus is 350 m^3 s^{-1} [33], with monthly averages ranging from 1 to 2200 m^3 s^{-1}. The Tagus estuary is a relatively shallow mesotidal system with semi-diurnal tidal regime (1 to 4 m in amplitude range). The surface area is about 320 km^2 and the mean volume 1900×10^6 m^3. Intertidal mudflats cover an area of about 20 to 40% of the estuary.

The coastal area off Tagus estuary is characterized by the presence of upwelling plumes originated by jet-like flow extending more than 20 km seaward [34]. Advection of warmer oligotrophic oceanic waters into the shelf occurs during autumn and winter when southerly winds dominate, intensifying the poleward flow [35]-[37]. Episodes of reverse winds can occur during both seasons. In the absence of coastal upwelling, the surface circulation is predominantly northward [36] as a result of the geostrophic equilibrium. Also, the plume of estuarine waters is highly influenced by the coastline geometry. Intense freshwater discharge events under highly variable wind direction conditions in winter and strong upwelling episodes in spring-summer as well as fortnightly spring-neap tidal cycle, affect strongly the shape and size of Tagus plume [38]. While the plume is usually trapped close to the shore and transports estuarine water northward along the coast, under persistent northern wind conditions the plume is displaced offshore.

A significant amount of phytoplankton is exported from Tagus to the estuarine plume. Field and modelling studies suggest that nutrients are not depleted by primary producers due to light limitation inside the estuary and end up by being exported, eventually enhancing primary production in the coastal area [39]-[43]. Tagus estuary is also a major source of nutrients [43] and suspended matter to the adjacent coastal area [38]. The transport and transformation of such materials in the area is regulated by the interplay of dynamics and structure of the Tagus plume and hydrological characteristics of the coastal area [44].

The Tagus estuary and adjacent coastal waters are sources of CO_2 to the atmosphere, with winter values ranging from 29 to 419 mmol C m^{-2} d^{-1} in the estuary, and up to 34 mmol C m^{-2} d^{-1} in the adjacent waters [45]. The carbonate system parameters have been evaluated in Tagus estuary and adjacent coastal area from 1999 to 2007 [46], and TA highest values (~4600 µmol kg^{-1}) were recorded in 2002 spring [26].

Sampling program

Surface water sampling was accomplished during ebb tide for a total of 16 stations distributed in the study area (see Figure 1), in two distinct winter periods (March 2001 and February 2004), defining winter as beginning in 21 December and ending in 21 March.

Parameters Determination

Temperature (T) and salinity (S; PSS-78) parameters were determined in situ with a Seabird SBE19/CTD (Conductivity - Temperature - Depth) probe. Salinity was calibrated with an AutoSal salinometer using IAPSO standard seawater, with a variation coefficient of 0.003%.

Dissolved oxygen (DO) was analysed following the Winkler method [47] using a whole-bottle manual titration, and the coefficient of variation associated with the method ranged from 0.08 to 0.25%. pH was measured immediately after sample collection at 25°C, using a Metrohm 704 pH-meter and a combination electrode (Metrohm) standardised against 2-amino-2-hydroxymethyl-1,3-propanediol seawater buffer (ionic strength of 0.7 M), at a precision of 0.005 pH units [48]. Total alkalinity (TA) samples were filtered through Whatman GF/F (0.7 µm) filters, fixed with $HgCl_2$ and stored (refrigerated not frozen) until use. Samples were then titrated automatically with HCl (~0.25 M HCl in a solution of 0.45 M NaCl) past the endpoint of 4.5 [48], with an accuracy of ±2 µmol kg^{-1}. The respectively accuracy was controlled against certified reference material supplied by A.G. Dickson (Scripps Institution of Oceanography, San Diego, USA). Discrete water samples were also taken for nutrient determination ($NO_3^- + NO_2^-$, referred as NO3; NH_4^+ referred as NH4; PO_4^{3-}, referred as PO4; $Si(OH)_4^-$, referred as Si (OH) 4), chlorophyll a (Chl a) and suspended particulate matter (SPM). Nutrient samples were filtered through MSI Acetate plus (0.45 µm) filters and analysed on a Traacs Autoanalyser, with a variation coefficient of ±1.0%. Chla was measured by filtering triplicate aliquots of 250 ml water through Whatman GF/F (0.7 µm) filters under a 0.2 atm vacuum, which were immediately frozen and later extracted in 90% acetone for analysis in a fluorometer Hitachi F-7000, calibrated with commercial solutions of Chl a(Sigma Chemical Co.). The coefficient of variation associated with the method was 1.8%. For SPM measurements six aliquots of 750–1000 ml water samples were filtered through pre-combusted (2 h at 450°C) Wathman GF/F (0.7 µm) filters and determined gravimetrically (drying at 70°C). A portable Vaisala® meteorological station (Datalogger Campbell Scientific CR510) coupled with a MetOne 034A anemometer located at 11 m height was used to measure in situ wind speed and direction data at 1-minute intervals at each station. Wind speed was referenced to a height of 10 m (u_{10}) using the algorithm given by Johnson [49]. We used one standard deviation of ±2 m s^{-1} as wind speed error.

Calculated Parameters

The upwelling indices (negative values indicate upwelling) were based on the northward wind stress component, and calculated according to Bakun [50]. Wind data was obtained from the meteorological weather station of Cape Carvoeiro located ~70 km north of Lisbon and supplied by the Portuguese Portuguese Institute for the Ocean and Atmosphere (IPMA, I.P.). Apparent oxygen utilisation (AOU) was calculated according to the equation:

$$AOU = O_{2sat} - DO \tag{1}$$

Where O_{2sat} is the oxygen saturation in equilibrium with atmosphere. pH values corrected to in situ temperature were calculated from total alkalinity (TA) and in situ pH and temperature following the procedure proposed by Hunter [51]. For these calculations the carbon dioxide constants of Millero et al. [52] were applied. The partial pressure of CO_2 in seawater (pCO_2) and the dissolved inorganic carbon (DIC) were calculated from the in situ temperature, TA and corrected pH, using the carbonic acid dissociation constants given by Millero et al. [52] and the CO_2 solubility coefficient of Weiss [53]. Errors associated with pCO_2 and DIC calculations were estimated to be ±10 µatm and ±5 µmol kg^{-1}, respectively (accumulated errors on TA and pH). The water-air CO_2 fluxes (CO_2 Flux) were computed according to the equation:

$$CO_2 Flux = k.K_0.\Delta pCO_2 \tag{2}$$

Where k is the gas transfer velocity (also referred to as piston velocity), $K0$ is the solubility coefficient of CO_2 and pCO_2 the water-air gradient of pCO_2. Positive fluxes indicate CO_2 upward water – air emission. The k value is based on the Wanninkhof [54] parameterization. Atmospheric CO_2 data were obtained from the Terceira Island's reference station (Azores, Portugal, 38°46'N 27°23'W), operated by the network of the National Oceanic and Atmospheric Administration (NOAA)/Climate Monitoring and Diagnostics Laboratory/Carbon Cycle Greenhouse Gases Group[55]. Subsequently, the observed atmospheric CO_2 content in mole fraction (in dry air) was converted into wet air values using the algorithms given by Dickson et al. [48]. Atmospheric pCO_2 data obtained from our single day shipboard were only available for some sampling periods, while Terceira data represent a readily accessible continuous thropospheric dataset for the complete study period. Significant correlations were found between Terceira data and shipboard data available ($r^2 = 0.910$, $p < 0.05$, n = 45). The discrepancies lie between 3 and 13 µatm, and the impact of using Terceira data on this study was considered negligible.

Statistical Analysis

Contour plots were created using Surfer 8.0® (Golden Software, 2002) following the kriging interpolation technique considering a linear interpolation with a slope of one. Exploratory analysis and statistical procedures were implemented using the statistical software Statistica 6.0® (Statsoft Inc., 2001). Differences between sampling periods in the measured/calculated physical-chemical and biological parameters were assessed using an analysis of variance (ANOVA), and differences between means have been considered statistically significant for $p < 0.05$.

Model Application

The Model

The MOHID Water Modelling System (www.mohid.com) was applied to this study to simulate the circulation regime of the study area. MOHID is a three-dimensional marine model that has been implemented in several studies of estuaries and shelf circulation [56]-[60]. MOHID employs a 3D finite-volume approach for spatial discretization [61] using an Arakawa-C grid [62] to perform the computations. For the baroclinic force, the MOHID system uses a z-level approach with a partial step approach [63]. Temporal discretization is performed by a semi-implicit ADI (Alternating Direction Implicit) algorithm with two time levels per iteration. The hydrodynamic governing equations are the momentum and the continuity equations. The hydrodynamic model solves the primitive equations in Cartesian coordinates for incompressible flows.

The momentum and mass evolution equations are:

$$\frac{\partial u_i}{\partial t} + \frac{\partial(u_i u_j)}{\partial x_j} = -\frac{1}{\rho_0}\frac{\partial p_{atm}}{\partial x_i} - g\frac{\rho(\eta)}{\rho_0}\frac{\partial \eta}{\partial x_i}$$
$$-\frac{g}{\rho_0}\int_{x_3}^{\eta}\frac{\partial p'}{\partial x_i}dx_3 + \frac{\partial}{\partial x_j}\left(v\frac{\partial u_i}{\partial x_j}\right) - 2\varepsilon_{ijk}\Omega_j u_k \quad (3)$$

$$\frac{\partial \eta}{\partial t} = -\frac{\partial}{\partial x_1}\int_{-h}^{\eta} u_1 dx_3 - \frac{\partial}{\partial x_2}\int_{-h}^{\eta} u_2 dx_3 \quad (4)$$

Where u_i is the velocity vector component in the Cartesian x_i directions, η is the free surface elevation, v is the turbulent viscosity and p_{atm} is the atmospheric pressure. ρ' is the density anomaly, ρ_0 is the reference density, g is the acceleration of gravity, t is the time, h is the depth, Ω is the Earth's velocity of rotation and ε is the alternate tensor.

The horizontal and vertical advection of momentum, heat and mass is computed using a Total Variation Diminishing (TVD) Superbee method [64]. Vertical turbulent viscosity/diffusivity coefficients are computed using a k-epsilon model coupling the MOHID system to the General Ocean Turbulence Model (GOTM) [65].

Modelled Scenarios

Two distinct winter episodes were modelled: March 2001 and February 2004. Both scenarios simulate oceanic conditions based on realistic forcing for river discharge (Figure 2) and wind conditions (Figure 3). We have adopted a method using a direct initialization with values from the MERCATOR solution [66]. This methodology interpolates the initial velocity field, temperature, salinity and sea surface height from the MERCATOR solution for the D2 grid assuming geostrophic balance. A two-month period was prescribed as a spin-up period.

Model Setup

The numerical model was implemented using a two level one-way nesting configuration. The first domain (D1) is a 2D barotropic tidal-driven model, forced only with the FES2004 (Finite Element Solution) tidal atlas [67], [68]. This domain covers most of the Atlantic coast of Iberia and Northwest Morocco, and has variable horizontal resolution (0.02°-0.04°). The second (D2) level is a 3D baroclinic model with a 0.02° horizontal resolution and includes the Tagus Promontory area. This domain is directly coupled to D1 at the open boundaries using a one-way downscaling to impose the solution of D1. For D1 low-frequency open boundary conditions for salinity, temperature and U and V velocity components are interpolated via a downscaling of the MERCATOR operational solution for the Northeast Atlantic area (Mercator-Océan Psy2V3). A z-level vertical discretization was adopted for D2 with 33 vertical layers. In this application we have set a time step of 60 s for D1 and 15 s for D2.

Hourly values for wind, air temperature, relative humidity, barometric pressure and downward longwave and shortwave radiation, were used to calculate air-sea heat and momentum fluxes using bulk formulae. The data for atmospheric forcing was retrieved from an atmospheric modelling system based on the MM5 (Mesoscale Meteorological Model 5) model running at IST (http://meteo.ist.utl.pt). For land boundary conditions, the model uses realistic freshwater discharge and a null mass and momentum flux is imposed. River outflow was prescribed using outflow values from the Portuguese Water Institute (INAG) gauges for Tagus River.

AUTHORS' CONTRIBUTIONS

APO conceived this study, contributed to all sections and coordinated the main writing process. APO and MM both analyzed the results and contributed equally to the manuscript. GC and RN provided valuable input for the data analysis and discussion sections. All authors have read and approved the final manuscript.

ACKNOWLEDGEMENTS

We acknowledge the captain and the crew of RV "Mestre Costeiro" and RV "Capricórnio" for their excellent support and cooperation. We are grateful to our colleagues António Correia, António Pereira, Célia Gonçalves, Conceição Araújo, Isaura Franco, Luís Palma Oliveira, Maria Rosa Pinto and Paula Cabeçadas for their sampling, technical and analytical assistance. Thanks are due to Marta Nogueira for CTD data acquisition. This work was funded by the European Commission, Programa POpesca MARE project 22-05-01-FDR-0015 and the Portuguese Science Foundation (FCT) with which A.P. Oliveira had a Ph.D. grant, and by the Project BioPlume - Dependence of coastal ecosystems on river run-off: today & tomorrow (PTDC/AAG-REC/2139/2012).

REFERENCES

1. Morris AW, Allen JI, Howland RJM, and Wood RG: The estuary plume zone: source or sink for land-derived nutrient discharges? Estuar Coast Shelf Sci 1995, 40:387-402.
2. de la Paz M, Gómez-Parra A, Forja J: Inorganic carbon dynamic and air–water CO_2 exchange in the Guadalquivir Estuary (SW Iberian Peninsula). J Mar Syst 2007, 68:265-77
3. Gaston TF, Schlacher TA, and Connolly RM: Flood discharges of a small river into open coastal waters: plume traits and material fate. Estuar Coast Shelf Sci 2006, 69:4-9.
4. Lohrenz SE, Fahnenstiel GL, Redalje DG, Lang GA, Dagg MJ, Whitledge TE, et al.: Nutrients, irradiance, and mixing as factors regulating primary production in coastal waters impacted by the Mississippi River plume. Cont Shelf Res 1999, 19:1113-41.

5. Sanders R, Jickells T, Mills D: Nutrients and chlorophyll at two sites in the Thames plume and southern North Sea. J Sea Res 2001, 46:13-28.
6. Dagg M, Benner R, Lohrenz S, and Lawrence D: Transformation of dissolved and particulate materials on continental shelves influenced by large rivers: plume processes. Cont Shelf Res 2004, 24:833-58.
7. Dagg MJ, Bianchi T, McKee B, and Powell R: Fates of dissolved and particulate materials from the Mississippi river immediately after discharge into the northern Gulf of Mexico, USA, during a period of low wind stress. Cont Shelf Res 2008, 28:1443-5
8. Dai M, Zhai W, Cai W-J, Callahan J, Huang B, Shang S, et al.: Effects of an estuarine plume-associated bloom on the carbonate system in the lower reaches of the Pearl River estuary and the coastal zone of the northern South China Sea. Cont Shelf Res 2008, 28:1416-23.
9. Borges AV, Frankignoulle M: Daily and seasonal variations of the partial pressure of CO_2 in surface seawater along Belgian and southern Dutch coastal areas. J Mar Syst 1999, 19:251-66.
10. Borges AV, Tilbrook B, Metzl N, Lenton A, Delille B: Inter-annual variability of the carbon dioxide oceanic sink south of Tasmania. Biogeosciences 2008, 5:141-55.
11. Brasse S, Nellen M, Seifert R, Michaelis W: The carbon dioxide system in the Elbe estuary. Biogeochemistry 2002, 59:25-40.
12. Schiettecatte L-S, Gazeau F, van der Zee C, Brion N, Borges AV: Time series of the partial pressure of carbon dioxide (2001–2004) and preliminary inorganic carbon budget in the Scheldt plume (Belgian coastal waters). Geochem Geophys Geosyst 2006., 7
13. Bozec Y, Cariou T, Mace E, Morin P, Thuillier D, Vernet M: Seasonal dynamics of air-sea CO_2 fluxes in the inner and outer Loire estuary (NW Europe). Estuar Coast Shelf Sci 2012, 100:58-71.

14. de la Paz M, Padin XA, Rios AF, Perez FF: Surface fCO_2 variability in the Loire plume and adjacent shelf waters: high spatio-temporal resolution study using ships of opportunity. Mar Chem 2010, 118:108-18.
15. Zhai W, Dai M: On the seasonal variation of air – sea CO_2 fluxes in the outer Changjiang (Yangtze River) Estuary, East China Sea. Mar Chem 2009, 117:2-10.
16. Kumar MD, Naqvi SWA, George MD, Jayakumar DA: A sink for atmospheric carbon dioxide in the northeast Indian Ocean. J Geophys Res 1996, 101:18121-5.
17. Bakker DCE, de Baar HJW, de Jong E: The dependence on temperature and salinity of dissolved inorganic carbon in East Atlantic surface waters. Mar Chem 1999, 65:263-80.
18. Chen C-TA, Wang S-L: Carbon, alkalinity and nutrient budgets on the East China Sea continental shelf. J Geophys Res 1999, 104:20675-86.
19. Cai W-J: Riverine inorganic carbon flux and rate of biological uptake in the Mississippi River plume. Geophys Res Lett 2003, 30:1032.
20. Körtzinger A: A significant CO_2 sink in the tropical Atlantic Ocean associated with the Amazon River plume. Geophys Res Lett 2003, 30:2287.
21. Lohrenz SE, Cai W-J: Satellite ocean color assessment of air-sea fluxes of CO_2 in a river-dominated coastal margin. Geophys Res Lett 2006., 33
22. Huang WJ, Cai WJ, Powell RT, Lohrenz SE, Wang Y, Jiang LQ, et al.: The stoichiometry of inorganic carbon and nutrient removal in \newline the Mississippi River plume and adjacent continental shelf. Biogeosciences 2012, 9:2781-92.
23. Borges A, Frankignoulle M: Distribution and air-water exchange of carbon dioxide in the Scheldt plume off the Belgian coast. Biogeochemistry 2002, 59:41-67.
24. Salisbury J, Vandemark D, Hunt C, Campbell J, Jonsson B, Mahadevan A, et al.: Episodic riverine influence on surface

DIC in the coastal Gulf of Maine. Estuar Coast Shelf Sci 2009, 82:108-18.
25. Cooley SR, Yager PL: Physical and biological contributions to the western tropical North Atlantic Ocean carbon sink formed by the Amazon River plume. J Geophys Res-Oceans 2006, 111:C08018.
26. Cabeçadas G, Oliveira AP: Impact of a Coccolithus braarudii bloom on the carbonate system of Portuguese coastal waters. J Nannoplankton Res 2005, 27:141-7.
27. Oliveira AP: Air-water CO2 fluxes in a Portuguese estuarine system and adjacent coastal waters (in Portuguese). PhD Thesis. Instituto Superior Técnico, Lisboa; 2011.
28. Dagg MJ, Breed GA: Biological effects of Mississippi River nitrogen on the northern Gulf of Mexico - a review and synthesis. J Mar Syst 2003, 43:133-52.
29. Borges AV, Delille B, Schiettecatte LS, Gazeau F, Abril G, Frankignoulle M: Gas transfer velocities of CO_2 in three European estuaries (Randers Fjord, Scheldt, and Thames). Limnol Oceanogr 2004, 49:1630-41.
30. Takahashi T, Olafsson J, Goddard JG, Chipman DW, Sutherland SC: Seasonal variation of CO_2 and nutrients in the high-latitude surface oceans - a comparative study. Global Biogeochem Cy 1993, 7:843-78.
31. Oliveira A, Fortunato AB, Rego JRL: Effect of morphological changes on the hydrodynamics and flushing properties of the Obidos lagoon (Portugal). Cont Shelf Res 2006, 26:917-42.
32. Gypens N, Lancelot C, and Borges AV: Carbon dynamics and CO_2 air-sea exchanges in the eutrophied coastal waters of the Southern Bight of the North Sea: a modelling study. Biogeosciences 2004, 1:147-57.
33. Santos FD, Forbes K, Moita R: Climate change in Portugal. Scenarios, Impacts and Adaptation Measures – SIAM Project. Lisbon, Portugal, Gradiva; 2002.
34. Moita M, Oliveira P, Mendes J, and Palma A: Distribution of chlorophyll a and gymnodinium catenatum associated with

coastal upwelling plumes off central portugal. Acta Oecol 2003, 24:125-32.
35. Fiúza AFG, Macedo ME, Guerreiro MR: Climatological space and time variation of the Portuguese coastal upwelling. Oceanol Acta 1982, 5:31-40.
36. Haynes R, Barton ED: A poleward flow along the Atlantic coast of the Iberian Peninsula. J Geophys Res-Oceans 1990, 95:11425-41.
37. Peliz A, Rosa TL, Santos AMP, Pissarra JL: Fronts, jets, and counter-flows in the Western Iberian upwelling system. J Mar Syst 2002, 35:61-77
38. Valente AS, da Silva JCB: On the observability of the fortnightly cycle of the Tagus estuary turbid plume using MODIS ocean colour images. J Mar Syst 2009, 75:131-7.
39. Mateus M, Leitão PC, de Pablo H, Neves R: Is it relevant to explicitly parameterize chlorophyll synthesis in marine ecological models? J Mar Syst 2012, 94:23-33.
40. Mateus M, Neves R: Evaluating light and nutrient limitation in the Tagus estuary using a process-oriented ecological model. Journal of Marine Engineering and Technology 2008, A12:43-54.
41. Mateus M, Vaz N, Neves R: A process-oriented model of pelagic biogeochemistry for marine systems. Part II: Application to a mesotidal estuary. J Mar Syst 2012, 94:90-101.
42. Saraiva S, Pina P, Martins F, Santos M, Braunschweig F, Neves R: Modelling the influence of nutrient loads on Portuguese estuaries. Hydrobiologia 2007, 587:5-18
43. Cabeçadas L, Brogueira , Cabeçadas G: Phytoplankton spring bloom in the Tagus coastal waters: hydrological and chemical conditions. Aquatic Ecology 1999, 33:243-250.
44. Oliveira PB, Nolasco R, Dubert J, Moita T, Peliz A: Surface temperature, chlorophyll and advection patterns during a summer upwelling event off central Portugal. Cont Shelf Res 2009, 29:759-74.

45. Oliveira AP, Nogueira M, Cabecadas G: CO_2 variability in surface coastal waters adjacent to the Tagus Estuary (Portugal). Cienc Mar 2006, 32:401-11.
46. Oliveira AP, Cabeçadas G, Pilar-Fonseca T: Iberia coastal ocean in the CO_2 sink/source context: Portugal case study. J Coast Res 2012, 28:184-95.
47. Carrit DE, Carpenter JH: Comparison and evaluation of currently employed modifications of the Winkler method for determining oxygen in seawater. A NASCO Report. J Mar Res 1966, 24:286-318.
48. Dickson AG, Sabine CL, Christian JR: Guide to best practices for ocean CO2 measurements. PICES Special Publication 3; 2007. p. 191.
49. Johnson HK: Simple expressions for correcting wind speed data for elevation. Coast Eng 1999, 36:263-9.
50. Bakun A: Coastal upwelling indices, west coast of North America. NOOA techn. Rep. NMFS-671. 1973.
51. Hunter KA: The temperature dependence of pH in surface seawater. Deep-Sea Res I Oceanogr Res Pap 1998, 45:1919-30.
52. Millero FJ, Graham TB, Huang F, Bustos-Serrano H, Pierrot D: Dissociation constants of carbonic acid in seawater as a function of salinity and temperature. Mar Chem 2006, 100:80-94.
53. Weiss RF: Carbon dioxide in water and seawater: the solubility of a non-ideal gas. Mar Chem 1974, 2:203-15.
54. Wanninkhof R: Relationship between wind speed and gas exchange over the ocean. J Geophys Res 1992, 97:7373-82.
55. Conway TJ, Lang PM, Masarie KA. Atmospheric carbon dioxide dry air mole fractions from the NOAA ESRL Carbon Cycle Cooperative Global Air Sampling Network, 1968-2011, Version: 2013-08-08, (ftp://ftp.cmdl.noaa.gov/ccg/co2/flask/event/), 2012.

56. Coelho HS, Neves RJ, Leitão PC, Martins H: A. S: The slope current along the Western European Margin: a numerical investigation. Bol Inst Esp Oceanogr 1999, 15:61-72.
57. Coelho HS, Neves RJ, White M, Leitao PC, Santos AJ: A model for ocean circulation on the Iberian coast. J Mar Syst 2002, 32:153-79.
58. Vaz N, Dias JM, Leitão PC, and Martins W: Horizontal patterns of water temperature and salinity in an estuarine tidal channel: Ria de Aveiro. Ocean Dyn 2005, 55:416-29.
59. Vaz N, Dias JM, Leitão PC, Nolasco R: Application of the Mohid-2D model to a mesotidal temperate coastal lagoon. Comput Geosci-Uk 2007, 33:1204-9.
60. Mateus M, Riflet G, Chambel P, Fernandes L, Fernandes R, Juliano M, et al.: An operational model for the West Iberian coast: products and services. Ocean Sci 2012, 8:713-32.
61. Martins F, Leitao P, Silva A, Neves R: 3D modelling in the Sado estuary using a new generic vertical discretization approach. Oceanol Acta 2001, 24:S51-62.
62. Arakawa A: Computational design for long-term numerical integration of the equations of fluid motion: Two-dimensional incompressible flow. Part I. J Comput Phys 1966, 1:119-43.
63. Kliem N, Pietrzak JD: On the pressure gradient error in sigma coordinate ocean models: a comparison with a laboratory experiment. J Geophys Res-Oceans 1999, 104:29781-99.
64. Vincent S, Caltagirone JP: Efficient solving method for unsteady incompressible interfacial flow problems. Int J Numer Methods Fluids 1999, 30:795-811.
65. Ruiz-Villarreal M, Bolding K, Burchard H, Demirov E: Coupling of the GOTM turbulence module to some three-dimensional ocean models. In Marine Turbulence: Theories, Observations, and Models Results of the CARTUM Project. Edited by Baumert HZ, Simpson JH, Sundermann J. Cambridge University Press, Cambridge; 2005:225-237.
66. Cailleau S, Chanut J, Levier B, Maraldi C, Reffray G: The new regional generation of Mercator Ocean system in the Iberian

Biscay Irish (IBI) area. Mercator Quarterly Newsletter 2010, 34:5-15.
67. Lyard F, Lefevre F, Letellier T, Francis O: Modelling the global ocean tides: modern insights from FES2004. Ocean Dyn 2006, 56:394-415.
68. Lefèvre F, Lyard FH, Le Provost C, Schrama EJO: FES99: a global tide finite element solution assimilating tide gauge and altimetric information. J Atmos Ocean Technol 2002, 19:1345-56

Citations

CHAPTER 1

K. S. C. Kuang, S. T. Quek, C. G. Koh, W. J. Cantwell, and P. J. Scully, "Plastic Optical Fibre Sensors for Structural Health Monitoring: A Review of Recent Progress," Journal of Sensors, vol. 2009, Article ID 312053, 13 pages, 2009. doi:10.1155/2009/312053.

CHAPTER 2

Westphalen, J. , M. Greaves, D. , Raby, A. , Hu, Z. , Causon, D. , Mingham, C. , Omidvar, P. , Stansby, P. and D. Rogers, B. (2014)

Investigation of Wave-Structure Interaction Using State of the Art CFD Techniques. Open Journal of Fluid Dynamics, 4, 18-43. doi: 10.4236/ojfd.2014.41003.

CHAPTER 3

Hasan I. Dawood, Kahtan S. Mohammed, and Mumtaz Y. Rajab, "Advantages of the Green Solid State FSW over the Conventional GMAW Process," Advances in Materials Science and Engineering, vol. 2014, Article ID 105713, 10 pages, 2014. doi:10.1155/2014/105713.

CHAPTER 4

Subir Paul, Anjan Pattanayak, and Sujit K. Guchhait, "Corrosion Behavior of Carbon Steel in Synthetically Produced Oil Field Seawater," International Journal of Metals, vol. 2014, Article ID 628505, 11 pages, 2014 doi:10.1155/2014/628505.

CHAPTER 5

Hana Bouteldja, Mohsen Hamidipour, Faïçal Larachi, Hydrodynamics of an inclined gas–liquid cocurrent upflow packed bed, Chemical Engineering Science, Volume 102, 11 October 2013, Pages 397-404, ISSN 0009-2509, http://dx.doi.org/10.1016/j.ces.2013.08.042.

CHAPTER 6

Zhongbin Ye, Guangfan Guo, Hong Chen, and Zheng Shu, "Interaction between Aqueous Solutions of Hydrophobically Associating Polyacrylamide and Dodecyl Dimethyl Betaine," Journal of Chemistry, vol. 2014, Article ID 932082, 8 pages, 2014. doi:10.1155/2014/932082.

CHAPTER 7

Chen, H., Stavinoha, S., Walker, M., Zhang, B. and Fuhlbrigge, T. (2014) Opportunities and Challenges of Robotics and Automation in Offshore Oil & Gas Industry. Intelligent Control and Automation, 5, 136-145. doi: 10.4236/ica.2014.53016.

CHAPTER 8

L. Wang and Q. Ni, "Vibration of Slender Structures Subjected to Axial Flow or Axially Towed in Quiescent Fluid," Advances in Acoustics and Vibration, vol. 2009, Article ID 432340, 19 pages, 2009. doi:10.1155/2009/432340.

CHAPTER 9

Ana P Oliveira, Marcos D Mateus, Graça Cabeçadas, and Ramiro Neves, Water-Air CO_2 Fluxes in the Tagus Estuary Plume (Portugal) during Two Distinct Winter Episodes, doi:10.1186/s13021-014-0012-3.

Index

A

Acrylamide (AM) 173, 174
Acrylic acid (AA) 173
Arbitrary-Lagrangian Eulerian (ALE) 49
Axial flow 219, 220, 221, 243, 245, 248, 250, 251, 253, 254, 256, 262, 263, 264, 265, 273, 274, 275, 277, 278

B

Bureau of Ocean Energy Management, Regulation and Enforcement (BOEMRE) 201

C

Clampedclamped 223
Computational fluid dynamics (CFD) 40, 153
Constant phase element (CPE) 128
Control-Volume Finite Element (CV-FE) 47
Critical association concentration (CAC) 187, 190
Cyclic transparent optical polymer (CYTOP) 4

D

Dissolved inorganic carbon (DIC) 283, 291, 296, 302
Dissolved oxygen (DO) 300
Dynamic interfacial tension (DIFT) 176, 181

E

Electrical capacitance tomography (ECT) 150, 155, 166
Electrochemical impedance spectroscopy (EIS) 128
Environmental, Health and Safety (EHS) 216

F

Fibre Bragg grating (FBG) 15, 22
Fibre Bragg gratings (FBG) 3
Fibre metal laminate (FML) 8
Finite Element (FE) 43
Finite Volume (FV) 43, 44
Fluidization heterogeneity 152, 153
Fowinduced 223, 266
Friction stir welding (FSW) 94

G

Gas metal arc welding (GMAW) 94
General Ocean Turbulence Model (GOTM) 304
Graded-index (GI) 20
Grid convergence index (GCI) 66

H

Heat affected zone (HAZ) 95, 109
High resolution interface capturing scheme (HRIC) 45
Hydrophobically associating polyacrylamide (HAPAM) 172, 173, 174, 175
Hydrophobically associating polymer (HAPAM) 190

I

Impulse response functions (IRF) 42
Inner diameter (ID) 154
Interfacial tension (IFT) 172, 176, 180

L

Liquefied petroleum gas (LPG) 200
Long period gratings (LPGs) 24

N

Natural Sciences and Engineering Research Council (NSERC) 167
Nugget zone (NZ) 95
Numerical wave tank (NWT) 44, 51, 76

O

Offshore oil 198, 199, 200, 201, 203, 206, 210
Optical time-domain reflectometry (OTDR) 18

P

Partially melted zone (PMZ) 109
Permanently unmanned area (PUA) 210
Plastic optical fibres (POFs) 3
Polymer optical fibre (POF) 22
Power-take-off (PTO) 41

Index 321

S

Saturated calomel electrode (SCE) 127
Shape memory alloy (SMA) 8
Slender structures 219, 220, 221, 253, 257, 262, 263, 264
Smoothed Particle hydrodynamics (SPH) 43
Smoothed Particle Hydrodynamics (SPH) 40, 48, 90
Straight pipes 222, 223, 226, 232, 233, 255
Structural health monitoring (SHM) 2
Suspended particulate matter (SPM) 301

T

Thermomechanically affected zone (TMAZ) 95
Total alkalinity (TA) 295, 296
Total Variation Diminishing (TVD) 304

V

Volume of fluid (VoF) 43

W

Wastewater 150
Wave energy converter (WEC) 40
Welding zone (WZ) 109

X

X-ray diffraction (XRD) 111